NAVIGATING

THE BRAIN

First published in 2024
by Riverside Press
an imprint of
UniPress Books Ltd
World's End Studios
London SW10 0RJ
United Kingdom

ISBN 978-1-7397988-7-1
E-book ISBN 978-1-7397988-8-8

British Library Cataloguing-in-Publication Data
A catalogue record for this book is available
from the British Library.

Publisher: Jason Hook
Project manager: Katie Crous
Design: Luke Herriott

10 9 8 7 6 5 4 3 2 1

Printed in China

unipressbooks.com

NAVIGATING

THE BRAIN

FIND YOUR WAY
THROUGH BIG IDEAS

RITA CARTER

RIVERSIDE PRESS

A century ago, the brain was, effectively, a black box. Things went in from the outside world – sights, sounds, words and so on – and behaviour came out. It was impossible to see what happened in the middle. Sigmund Freud had some brilliant ideas about what was going on and made them into the massively successful psychoanalysis. Then behaviourism developed, putting psychology on a more rigorously scientific footing. Although informative and still useful, behaviourism neglects what most of us think is the most important thing about the mind: our subjective universe.

In the late 20th century, technology prised open the black box of the brain and, at last, made it possible to look inside. One by one, innovative imaging techniques started to reveal the complex machinations that give rise to the human mind. *Navigating the Brain* is designed to explore how this new view of the brain unites psychology with its physiological foundation.

We start by addressing the relationship between brain and mind, the nature of self, and consider how the human brain became like it is and how it differs from the brains of other species. We then look at the physical brain: its gross structure and the minute elements that make it work. We explain how such different aspects of humanity – base instinct, emotion, sensation and high-level

TION

thought, creativity and judgement – can emerge from a single organ.

Moving on, the emphasis shifts to the way the brain works. Most of us know that cognition is reliant on the flow of electrical information between neurons, but how is it generated, how does it move and how does it become sights, sounds and feelings? We then consider memory and learning, explaining what memories are and how they are stored, brought to mind, changed, hidden and forgotten.

Next in our journey is a tour through emotions – what they are, why we have them and how they differ between individuals. We examine the evidence for animal emotion and the emotional difference between men and women. Continuing with the theme of difference, we assess the brain's contribution to the massive diversity of human life. It explains how microscopic differences in brain anatomy produce a near-infinite range of personalities, experience and behaviour.

We then move on to brain health and how best to keep this extraordinary part of your body working well throughout your life. We venture beyond the brain, ending with the most mysterious and intriguing subject in the universe: consciousness. What is it exactly, and what is its connection to the brain? Finally, we dare to ask if artificial intelligence will ever be able to develop consciousness too?

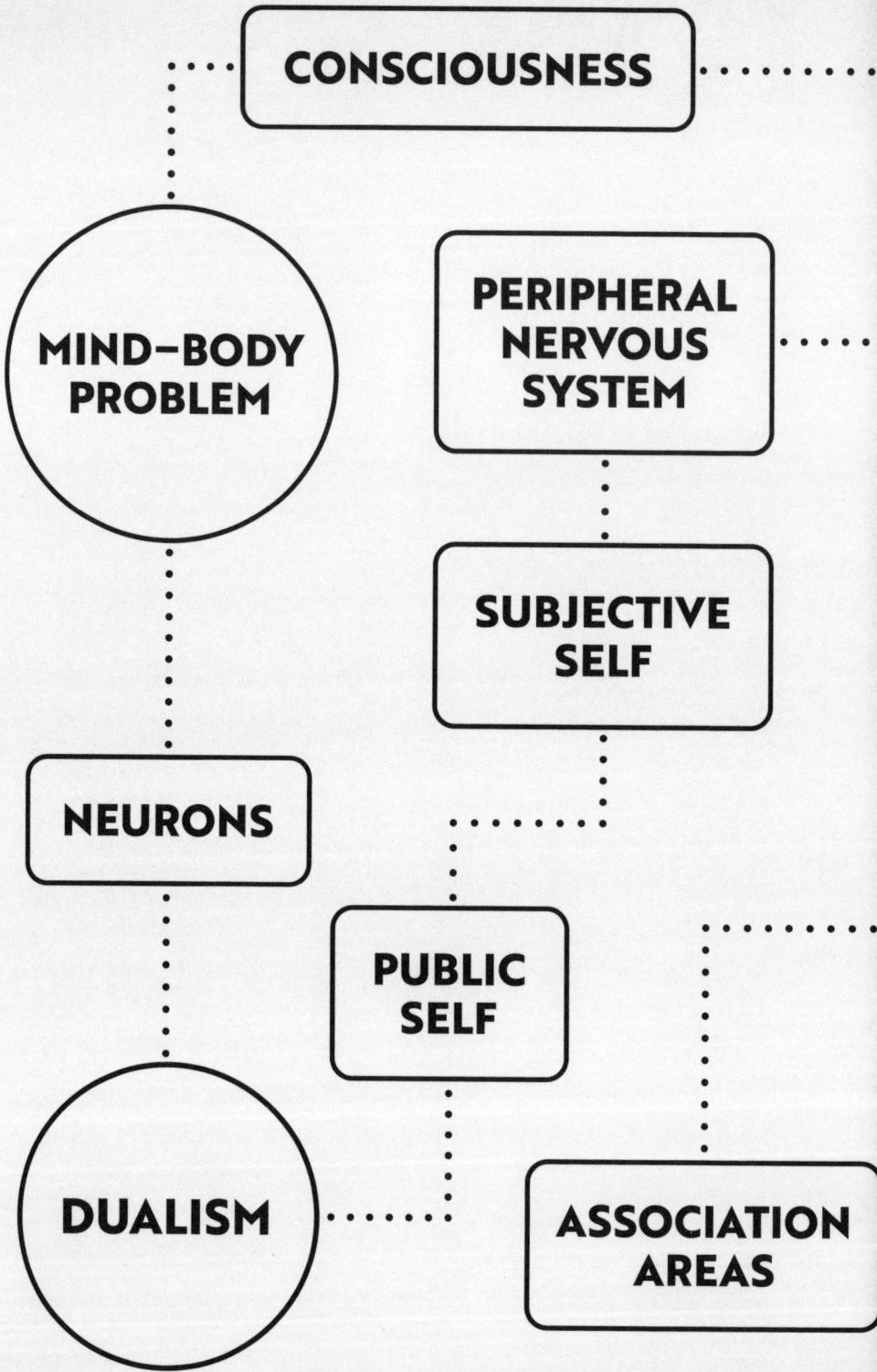

CHAPTER 1

BRAIN & MIND

ANIMALS

LANGUAGE

EVOLUTION

GROWTH

INTRODUCTION

The brain is unlike any other organ. Your heart, liver and kidneys are amazing, but what they do is fairly easy to understand now that biologists have worked it out. The brain, by contrast, presents what seem at first to be impenetrable mysteries. Instead of producing chemicals or acting directly on other parts of the body, it seems to be responsible for the sensations, emotions, thoughts and perceptions that give meaning to our world. We have learned an enormous amount about how it does all this, but huge questions remain. How does the brain relate to the idea of the **MIND**? Is it one and the same as our subjective universe or does it somehow generate our universe or even 'tune in' to it? What about your 'self' – not just your physical body, or **PUBLIC SELF**, but the 'you' that can make decisions and judgements and that seems to have the freedom to act as it desires, your **SUBJECTIVE SELF**? How can that come from a 1.3-kilogram (3-pound) lump of flesh?

These questions do not have to bother you – a person can get by in life perfectly well without worrying about such things. But if you do think about them, you'll find yourself grappling with – and hopefully grabbed by – the greatest enigma in the world.

This chapter introduces you to the strange world of the human brain. It describes some of the evidence science has brought to bear on issues that philosophers pondered for centuries before tools were invented that made it possible to explore them methodically. We now know, for example, how infants' brains prune the excessive neuronal tissue they are born with to make sense of the world around them, sculpting recognisable sights and sounds out of the 'buzzing blooming confusion' of early **CONSCIOUSNESS**, as the great 19th-century US psychologist William James described it.

BRAIN IMAGING has correlated brain anatomy and behaviour to the extent that we can see how the anatomical differences between human brains and those of other species give rise to the differences we observe in the way animals behave. Advances in evolutionary science have given us an idea of how, when and why we diverged from our primate ancestors to become human, and how our brains drove changes in our behaviour that have come to make us (for good or bad) the species that dominates Earth.

Science has not yet provided us with all the answers, though, and intriguing questions remain. Would it be possible, for example, for the brain to survive if it were separated from the rest of the body and fed it a completely false world, Matrix-esque? As scientists produce brain-like **ORGANOIDS**, we might be on the brink of finding out.

BRAIN & MIND MAP

NEURAL INFORMATION
The continuous flow of electricity through the brain, most of which is concerned with unconscious processes, some of which produces conscious experience.

DUALISM
Relating to the mind–body problem, a collection of theories positing that mental phenomena are non-physical. First proposed by Descartes as a version known as Cartesian dualism.

BRAIN IMAGING
Technology that scans the brain for function and health; includes functional magnetic resonance imaging (fMRI), computerised tomography (CT), positron emission tomography (PET), electroencephalography (EEG) and magnetoencephalography (MEG).

GLIAL CELLS
Non-electric brain-tissue cells that structurally support, insulate and influence neurons.

NEURONS
Brain-tissue nerve cells that generate electricity and pass it on to others. There are about 90 billion neurons in the human brain.

AXON
Long tendril that snakes out from the body of a neuron and connects with other neurons.

MYELIN
A fatty substance that forms around axons as they mature, speeding and smoothing the electrical current that carries information between neurons.

BRAIN ORGANOIDS
Self-organising, three-dimensional tissue grown in laboratories from stem cells and capable, to a point, of mimicking normal brain activity.

PERIPHERAL NERVOUS SYSTEM
The nerves that extend from the brain to the rest of the body. The brain sends information to and around the system and receives information back.

READINESS POTENTIAL
Unconscious wave of neural activity that precedes a voluntary movement by about half a second. Usually interpreted as marking the start of preparation to move.

BRAIN

MIND

MIND–BODY PROBLEM
Also known as the 'hard problem', as coined by philosopher David Chalmers in 1994. The unresolved question of how flesh (the physical) and mind (the mental) are connected.

MIND
The seat of consciousness, our essence of being and the sum of our cognitive processes.

FRONTAL LOBES
The largest lobes in the brain, located behind the forehead; responsible for high-level functioning and vital to consciousness.

ASSOCIATION AREAS
Patches of cortex where small units of neural information (e.g. size, shape and colour of an object) are combined to create a larger perceptual component.

PUBLIC SELF
Physical existence as a body that other people can identify, structured by social context.

SELF

SUBJECTIVE SELF
Mental existence in the sense of being inside your body and your mind, with a sense that you decide, initiate and control your actions.

CONSCIOUSNESS
Our normal waking state, including perceptions, thoughts, feelings and awareness of the outside world and of self.

What's the difference between brain and mind?

⟶ The brain is a lump of body tissue in our head. The mind is what the lump does: perceives, thinks, feels, directs our actions and sometimes generates consciousness.

Most of our organs do visible jobs such as pumping blood around or making recognisable objects like bile or faeces. The brain is different – what emerges from it seems to be a different type of thing altogether: 'spirit stuff', rather than anything in the material world.

The question of how flesh and mind are connected is known as the 'mind–body problem' and it has been debated for centuries. Initially, the brain did not even figure in many of these debates – ancient Egyptians removed it before sending a dead person off to the afterlife because, unlike other organs, they thought the brain would not be required. Many cultures have identified the heart, rather than the brain, as the seat of a person's soul.

The mind–body problem remains unsolved, but scientific research has established beyond reasonable doubt that the brain is responsible in some way for our actions and experiences. Brain imaging technology can even show where, and to some extent how, it all happens.

Brain tissue is unlike other flesh in that much of it is made of cells, called neurons, which generate electricity and pass it on to others. There are about 90 billion neurons in the human brain, and an even greater number of non-electric glial cells, which structurally support, insulate and influence neurons. Live neurons generate electricity by 'firing', and the strength, pattern, route and location of these pulses determine what the brain does – thinks or perceives something, produces an emotion or instructs the body to move.

The continuous flow of electricity through the brain is referred to as neural 'information'. Most of it is concerned with unconscious processes, such as maintaining heart rate and breathing, or handling the huge amount of backstage cognition that produces a perception or a spoken word. Some of it produces conscious experience – probably the most extraordinary thing in the universe.

DUALISM

Consciousness is so unlike anything in the natural world that many people intuitively believe it is some kind of 'other stuff'. The French philosopher René Descartes (1596–1650) formalised this belief with his theory of dualism – that mind and brain are distinct. He proposed that the pineal gland – a tiny brain nucleus now known to be involved in light/dark detection – was a sort of aerial, picking up spiritual vibes and moving the brain to act on them.

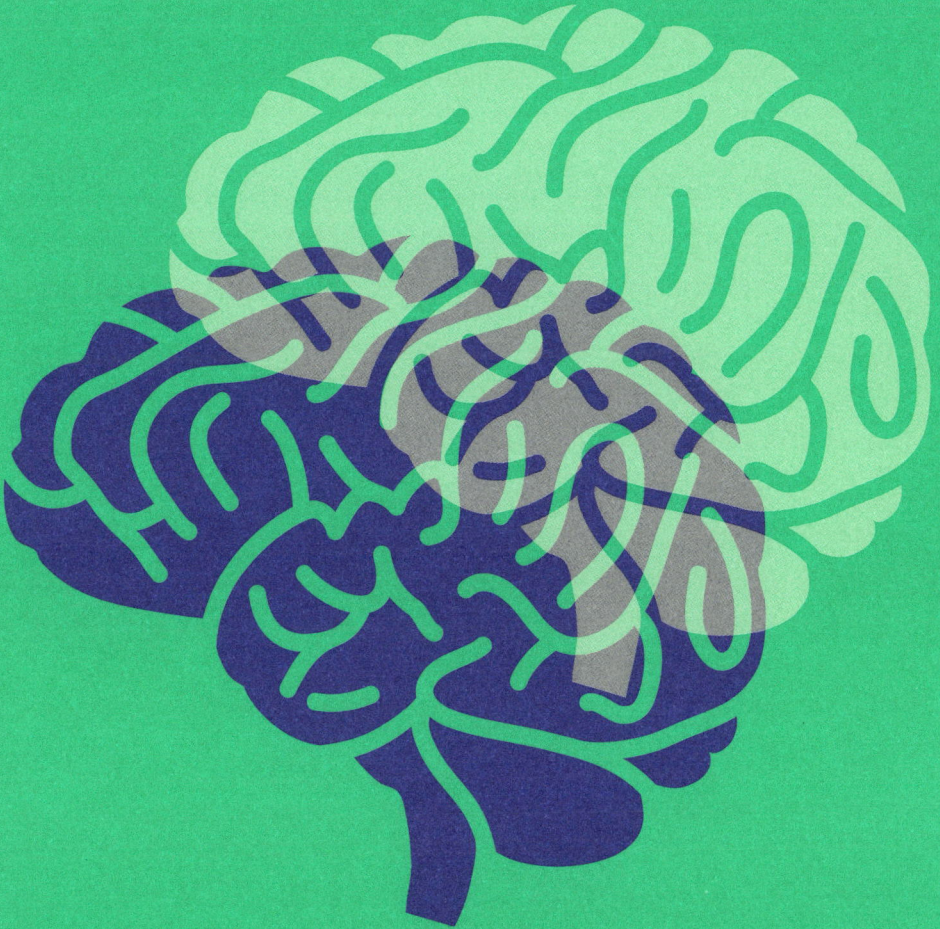

Am 'I' my brain?

→ It's probably the brain's cleverest trick – creating a sense of self that seems to be separate from your body. Don't be fooled.

It's easy to take your self for granted. When you see or feel something, you know that the thought or feeling is yours. Similarly, when you do something voluntarily – pick up a cup, for example – you assume that the act was initiated by you.

Yet this sense of ownership and agency is not a given. You exist, of course, as a body that other people can identify. This is your public self. Your subjective self, what's left if you take away memory and social context, is more complicated. It has two main components. One is ownership: the sense of being inside your body and your mind. You know what's part of you and what is not, and you identify with the sensations, thoughts, perceptions and emotions that are yours. The second component is agency: the sense that you decide, initiate and control your actions.

These senses are operating, unconsciously, throughout our waking life. They disappear only if normal brain activity is disrupted, as it is sometimes in schizophrenia, with certain types of delusion and under the influence of some drugs. The effects experienced in these conditions demonstrate the importance of a continuous sense of self. Loss of ownership may result in a person believing that their thoughts are being spoken by outside voices, or they may fail to acknowledge their feelings or perceptions. They may even 'lose' their body, making them feel terrifyingly incomplete.

Agency is the feeling that we can act at will. It is a powerful conviction, but one that may be a delusion. In 1982 US neuroscientist Benjamin Libet (1916–2007) demonstrated that the unconscious brain activity that kicks off a 'voluntary' action occurs before a person decides to do it. The implication is that our brains decide our actions, and our conscious 'decision' is just something we experience as it happens.

READINESS POTENTIAL

An unconscious wave of activity called the Readiness Potential (RP) sweeps through the brain about half a second before a voluntary movement. It marks the start of the brainwork that has to be done before the act can happen.

Experiments show that people 'decide' to move just before the RP reaches a peak, and a split second later the movement occurs. One interpretation is that the 'decision' to move is just early recognition of what the brain is about to do anyway.

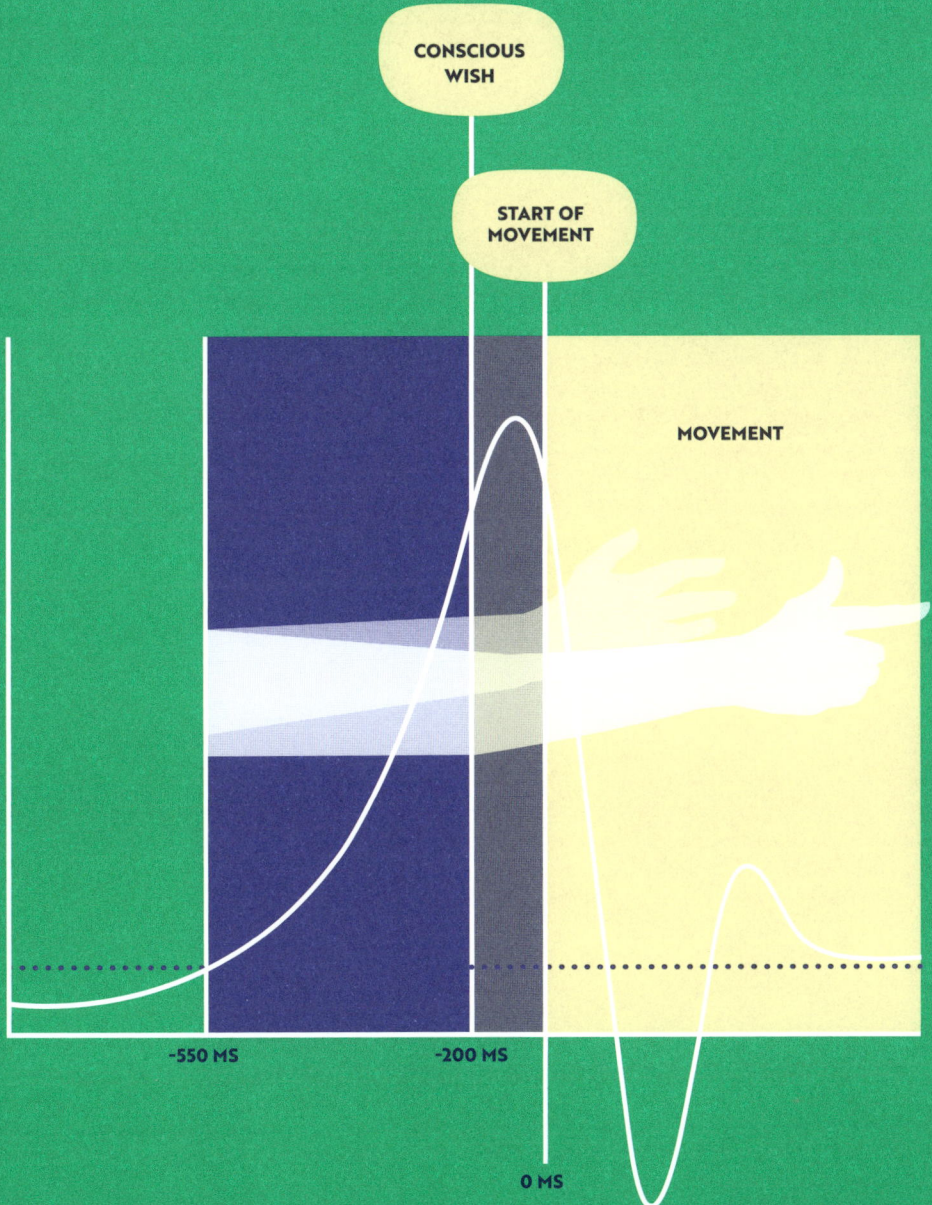

CONSCIOUS WISH

START OF MOVEMENT

MOVEMENT

-550 MS

-200 MS

0 MS

Could my brain work without my body?

→ Almost certainly not. Your brain is part of your body, so it is a bit like asking if your hand would keep working if you chopped it off.

We think of the brain as an organ that stops at the neck. The boundary is drawn mainly for convenience – there is no physical dislocation between the brain, spinal cord and peripheral nervous system, which sends nerves snaking through every part of the body to the tips of your toes. The brain sends information to and around the peripheral system and receives information back. If this two-way traffic is stopped, the brain knows nothing, and can do nothing.

Apart from that, the brain also requires a constant flow of blood to give it the oxygen and nutrients it needs to work. Depriving it of blood kills it almost instantly, which is why beheading is a very efficient way of killing someone.

A famous thought experiment, 'the brain in a vat', is used to explore what we can know to be true. You are asked to imagine a brain that some mad scientist has taken out of a body but kept alive in a vat by feeding it

all the oxygen, nutrients and hormones it would normally receive. It is also receiving transmissions of some sort that cause the brain to create sensations and perceptions – an entire virtual world. The brain could go for a walk in the park, and nothing about the experience would tell it that it wasn't feeling real grass beneath its real feet. The film *The Matrix* is a modern take on the idea. Given that the vat-brain's subjective experience is identical to how it would be if it was back in a skull, the question raised by the thought experiment is whether we – snug in our own skulls – can know if our experiences are real.

Philosophical questions aside, would it be possible, technically, to keep a human brain alive in a vat? For obvious ethical reasons, no one is known to have tried. However, researchers have maintained function in an isolated pig's brain for several hours by hooking it up to a machine that kept the cells fed and oxygenated, so perhaps it is only a matter of time.

BRAIN IN A VAT

In 2019 researchers kept an isolated pig's brain alive for six hours by perfusing it with oxygen and other nutrients – almost exactly the scenario imagined in the 'brain in a vat' thought experiment. The research was aimed at finding ways to help people survive brain injuries. Although the brain maintained many normal physiological processes, the researchers reported that there was no sign of the sort of neural activity associated with perception, sensation or consciousness.

Are human and animal brains different?

→ Yes. Some species don't have a brain at all – instead their nervous system is spread throughout the body. In vertebrates, though, the basic layout is common to all. The difference is in the wiring.

At first sight, the human brain looks much like that of any other large verterbrate. Certainly it's big – three times as big as the brain of our nearest relatives, chimpanzees. Yet it's not as big as that of, say, a dolphin or an elephant, and in relation to total body size it's comparable to the brain of a mouse.

In shape, the human brain's most notable distinction is the bulging frontal lobes. In cross-section, you see that the cortex – the wrinkly, grey skin – is layered, like a laminated windscreen. Most of the cortex has six welded sheets of tissue, known as neocortex. In other vertebrates, the cortex is thinner and more of it is an evolutionarily older type with fewer layers.

Beneath the cortex is a chunk of whitish tissue made of fat-sheathed tendrils (axons and dendrites) that snake between cortical neurons. This is the stuff that carries the electrical information that makes brains work. It is denser in humans, especially in the frontal lobes.

It is not simply neuronal volume that makes our brains so clever, however. It has more to do with the way they work together, which you can see through brain imaging technology such as fMRI.

The faculty that most distinguishes us from other animals is language. The brain areas that deal with this are situated (in most people) in the left hemisphere, around and in front of the ear. Oddly, these crucial parts can barely be discerned, even under a microscope. Rather than arising simply from growth, they are the result of massive internal rewiring that occurred around the time humans first stood up and started hunting. A neural pathway was formed between brain areas involved in gestures, vocalisation and hearing, and that pathway merged into those serving memory and recognition.

As well as speech, this new super-connectivity gave humans the ability to attach words to objects, creating symbols – the basis of abstract thought. Many other animal brains have brain areas dedicated to communication, but ours is uniquely interconnected with the parts that do our thinking.

ASSOCIATION AREAS

The brains of all mammals have 'association areas' – patches of cortex that combine current perceptions with memories, thoughts and muscle control in order to generate an appropriate response. A dog's brain will bind the sight of a running rabbit with the notion of food and the movements required to get it. Conclusion: it chases. Human association areas bring much more to the mix: ideas of pet rabbits, rabbit traps, Easter bunnies, cartoon characters, pests that eat crops – our potential responses are much wider and more flexible. It's what we call intelligence.

How did the human brain get like this?

→ Getting up on our hind legs kick-started the long evolutionary process that changed a small, sense-heavy ape brain into the thinking, talking, intelligent organ we carry today.

The first animals that were indisputably human were named for their characteristic stance: *Homo erectus*. Standing up produced both advantages and disadvantages for our ancestors. One advantage of standing tall is that it affords a wider view of the landscape, which has more visual components than a nose-to-ground view.

To make sense of all this new input, human brains got bigger in patches of cortex called 'association areas' – the places where details are combined to make the 'big picture'. The bigger the association areas, the more information a brain can deal with. Human frontal lobes are the biggest association areas of all.

Another advantage of being upright was the ability to see things at a distance, including potential prey. This made hunting easier, tipping humans away from a vegetarian diet. Animal protein provided extra calories for brain growth (the organ uses 20 per cent of our available energy) and hunting demands sleuth, timing and tool-making – skills that further stimulated frontal lobe expansion.

Hunting may also have helped to bring about speech. Footprints found in Africa suggest that people were hunting in teams 1.5 million years ago, and teamwork requires sophisticated communication. Hand gestures became increasingly complex, and neural links developed between the areas that control them and those, nearby, that control mouth movements, then breath. This made it possible to articulate vocal sounds, while linking them with the meaning inherent in gestures.

Standing up also brought a serious disadvantage: it narrowed the pelvis and made childbirth painful and difficult. The evolutionary fix was to have human infants born while their heads were still small and malleable. This allowed the skull to squeeze through the narrow birth canal, and also to expand to allow the brain to grow. Immature birth, however, meant human infants were dependent on their mothers for much longer than other species, and this in turn made females vulnerable because they were tied to their infants. One solution was to build complex social support networks. Those with good communication and social skills achieved this best and survived, along with the genes that gave them those abilities.

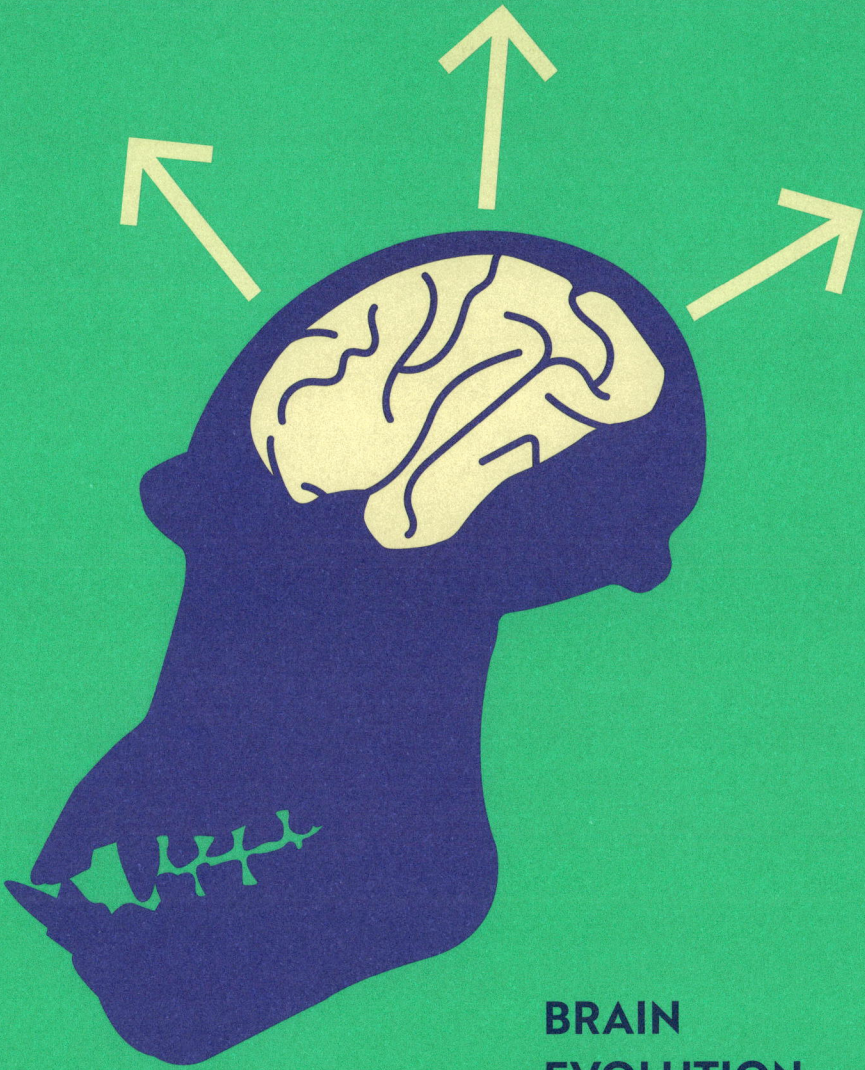

BRAIN EVOLUTION

The size of the hominid brain increased gradually and steadily over time until today it is three times the size of our very early ancestors. The gradual – rather than step-by-step – expansion was discovered by researchers at the University of Chicago who studied fossils from thirteen human species, starting with Australopithecus, 3.8 million years ago. Brains seemed to expand both within species and by the introduction of bigger-brained species.

Do our brains keep growing?

→ New brain cells are made by the million for a short period after birth, but neuronal growth calms down after about two years. The brain continues to become heavier and bigger, however, because the cells expand.

Our brains keep changing, but after infancy what grows mainly is the connective tissue between neurons rather than new cells. Long tendrils called axons snake out from the body of the neuron and seek out other neurons to connect with. As they mature, axons are wrapped in myelin, a fatty substance that works like insulation around a wire, speeding and smoothing the way of the electrical current that carries information from one neuron to others.

By the age of three these links have formed a dense network around which electrical information whizzes chaotically. The 'noise' in a toddler's brain is reduced over the next few years by a process called apoptosis – a natural dying away of neural connections that do not carry useful information. It is rather like a young shrub being pruned to encourage it to grow into shape.

What is left is those connections that are useful, such as those that carry information to the brain about where the foot is, so the child doesn't fall over when they walk. These continue to grow as more neurons send their axons along the same route. A threadlike connection between neurons that respond to a caregiver's voice saying 'no' and those that stop an action becomes, in time, a thick, and more effective, pathway.

Myelin growth continues until 30 years of age, and new connections form for as long as we learn new things, making the brain increasingly dense. New neurons continue to be made too, but only in a couple of areas; they do not contribute much to the overall growth of the brain. Most people's brains start to shrink after the age of about 60, even in the absence of dementia. On average, the weight of a brain when a person dies is 10 per cent less than when they were in early adulthood.

BABY BRAIN

Newborn babies have as many, or more, neurons as they will ever have, yet their brains expand dramatically in the first two years of life. Most of this growth is due to the creation of neuronal links, formed by cell axons stretching out and joining up with others. Many of these connections are not useful, however, and during the next few years they are pruned away, leaving a streamlined network of fewer but stronger links.

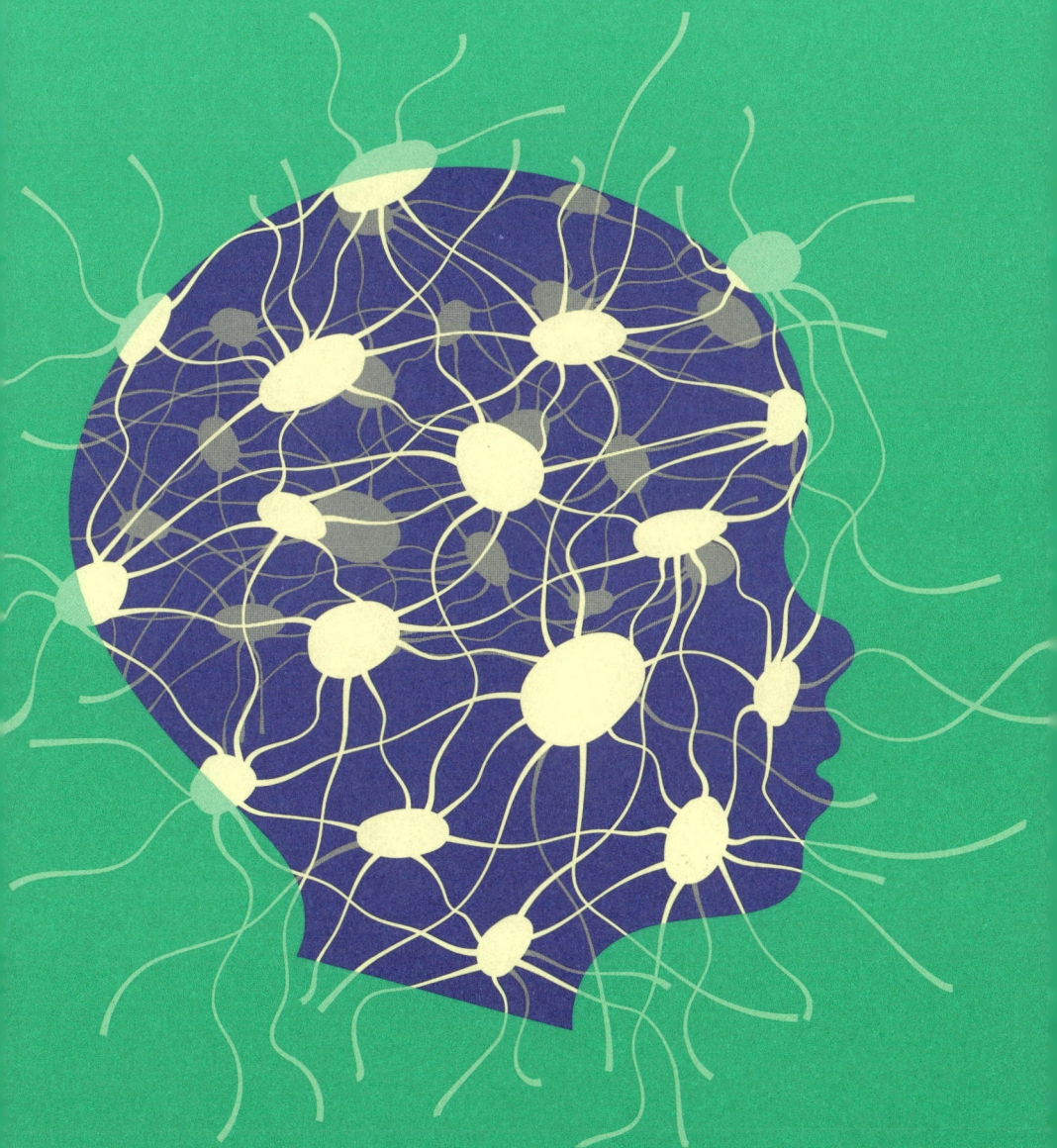

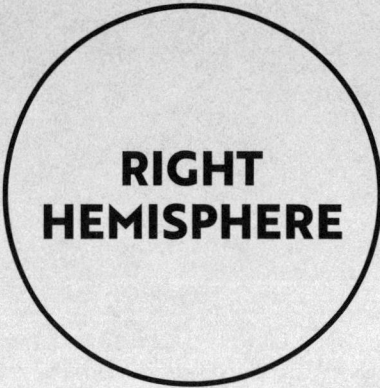

RIGHT HEMISPHERE

LEFT HEMISPHERE

BRAINSTEM

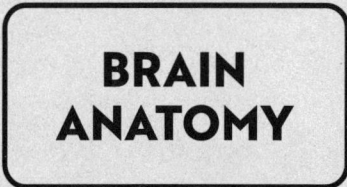

BRAIN ANATOMY

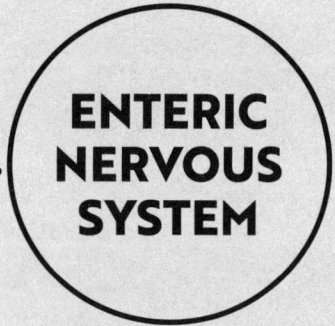

ENTERIC NERVOUS SYSTEM

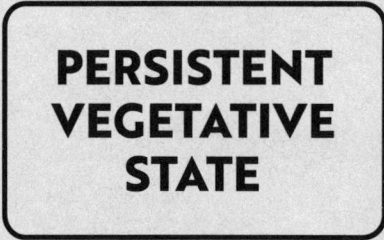

PERSISTENT VEGETATIVE STATE

BRAINSTEM DEATH

IONS

CHAPTER 2

BRAIN STRUCTURE

TRIUNE BRAIN

LONGITUDINAL FISSURE

LOBES

NEUROTRANSMITTERS

INTRODUCTION

Human behaviour is so varied and extraordinary it is easy to forget that it is produced by a physical organ, and not even a particularly interesting-looking one at that. To the naked eye, the brain is a greyish lump of wrinkled flesh, with nothing obvious to suggest it does anything special. Most early anatomists were more interested in the heart, which could be seen to pump life-giving blood and even respond to events and feelings – thumping faster when a person was excited, slower when relaxed. Not surprisingly, it was the heart, therefore, that was thought to be the seat of human emotion.

It took a very long time to get even a rough idea of which parts of the brain did what. It only became possible to look at a working brain with the invention of brain imaging technology in the late 20th century, so progress depended mainly on 'natural experiments': accidents that damaged particular areas of the brain, causing behavioural changes. By observing these changes and the corresponding damage location, scientists were able to start to compile a map of brain functions. The development of high-definition microscopes showed that, despite its appearance to the naked eye as undifferentiated flesh, the brain is actually composed of hundreds of distinct areas and fuelled by a special type of cell, the neuron. Technology is now revealing how neuronal activity in different areas produces our experience and behaviour.

This chapter describes some of what we now know about the surprisingly complicated structure of the human brain: the so-called 'reptile brain' that lurks at its base, with Freud's unconscious mind embodied in the **LIMBIC SYSTEM**, above it; the expanse of wrinkly grey skin – the **CORTEX** – with its multi-layered structure and two-way connections to the parts below. We also look at two widespread but often misunderstood ideas. One is that we have a 'second brain' in our guts, otherwise known as the **ENTERIC NERVOUS SYSTEM**, which influences our moods and mental health. The second is the idea that our **RIGHT HEMISPHERE** is creative while the **LEFT HEMISPHERE** is logical. In each case there is a scientific basis to these ideas, but the mythic dimensions they have achieved through repetition and simplification have distorted the physiological facts from which they emerged.

We also look at life itself: how neurons generate the electric energy that keeps us going, and whether there is any life at all without the brain.

BRAIN STRUCTURE MAP

HEMISPHERES

PAUL D. MACLEAN (1913–2007)
US neuroscientist who proposed the idea of the triune brain: that humans have three separate brains (reptile, mammalian and human) which have evolved on top of one another.

LEFT HEMISPHERE
Dominant half of the brain in most people; controls the right side of the body and does the most obvious language tasks; more concerned than the right hemisphere with details and plans.

RIGHT HEMISPHERE
Controls the left side of the body; detects speech rhythm and emotional tone; keeps a grasp on the wholeness of a situation; more visual and intuitive in most people than the left hemisphere.

LIMBIC SYSTEM
Part of the brain beneath the cortex that generates emotion, and is involved in learning, memory and many aspects of behaviour.

LOBES
Each hemisphere of the brain is divided into four lobes – frontal, parietal, temporal and occipital – each of which is associated with different functions.

BRAINSTEM
At the bottom of the brain, this part controls basic survival mechanisms, including breathing, blood pressure and sleep/wake cycles.

ENTERIC NERVOUS SYSTEM
Dense web of neurons embedded in the wall of the gastrointestinal system, stretching from stomach to rectum and affecting emotions, health and well-being.

CEREBRAL CORTEX
The wrinkled outer layer of the brain responsible for producing conscious cognition and high-level processes such as reasoning and problem-solving.

SYSTEMS

ANATOMY

NEUROTRANSMITTERS
Chemicals made and stored in the tips of axons, released when the axon fires and that carry electrical signals to the next neuron. There are more than a hundred different ones.

CEREBRUM
The big round part of the brain, divided into two hemispheres. The back part deals mainly with sensory perception, and the front makes sense of it.

SYNAPSES
Tiny gaps that separate axons from dendrites; the communication hub of neurons.

CEREBELLUM
Part of the brain that manages posture and balance and contributes to fine and automatic movements. It has some influence on emotions.

DENDRITES
Branch-like fibres with receptors on their ends that receive the electricity from the axons of neurons.

HYPOTHALAMUS
Cluster of tiny nuclei that regulate the state of the body, including appetite, urges, thirst and temperature.

PERSISTENT VEGETATIVE STATE
When a person survives severe damage to the brain yet continues to have functions such as breathing, circulation and sleep/wake cycles.

AMYGDALA
Almond-shaped nucleus in the limbic system which samples incoming information and generates emotional responses.

THALAMUS
In the centre of the brain, this part takes eveything in from the outside world (except smell) and directs it for further processing. Plays a crucial role in cognition.

HIPPOCAMPUS
Part of the limbic system that encodes, stores and retreives information. The seat of learning and memory.

Have we got three brains: reptile, mammal and human?

→ We only get one brain each. The 'triune' brain has some evolutionary justification, though, because we have three levels of cognition – physical survival, emotion, and conscious perception and thought – which occur mainly in three different brain layers.

The triune brain, proposed in the 1950s by US neuroscientist Paul D. MacLean (1913–2007), is the idea that humans have three separate brains which have evolved with one on top of the other. Although they are not separate, as MacLean seemed to suggest, there are three identifiable layers in a human brain, which emerged one by one as we evolved from reptiles, through mammals, to humans.

The triune brain theory has been heavily criticised for not accurately reflecting the way that all parts of the brain work together. It also looks doubtful now that evolution 'built' the human brain in the way MacLean supposed. Rather than having different overall structures, all vertebrates have the three brains MacLean identified, but in different proportions. Even reptiles have a part of the brain that is structured like the mammalian cortex. The idea of the triune brain idea has stuck, though, and remains useful as a way to distinguish the different types of cognition at each level of the brain.

MacLean called the lower part of the brain the 'reptile' or 'lizard' brain. It comprises the brainstem, which controls basic survival mechanisms like heartbeat and breathing, and the cerebellum, which manages posture, balance and automatic movements. The middle 'mammalian' part, which MacLean was first to label the limbic system, includes the areas concerned with memory and emotions; and the 'human' brain is the neocortex, where thinking, judging and talking happen.

The simple separation of these types of cognition is misleading because the three brains are melded and rarely work in isolation. In humans, for example, the cerebellum is involved in emotional regulation, inhibiting impulsive decisions, and even in some aspects of memory. The limbic system deals with much more than emotion – learning and memory is centred here, along with many aspects of behaviour. The cortex is in constant communication with the lower levels, and each affects the others.

THE TRIUNE BRAIN

The brain evolved like a fast-growing city, with new bits added on top. At its base, dating right back to reptiles, lie mechanisms that support breathing, blood pressure, attention and reflex action. Mammals evolved a second layer, now called the limbic system, which allowed them to make memories and emotions. The cortex evolved on top of that and grew extra layers to become the neocortex, which supports very complex behaviour. The neocortex then expanded to form the huge prefrontal area that distinguishes human brains.

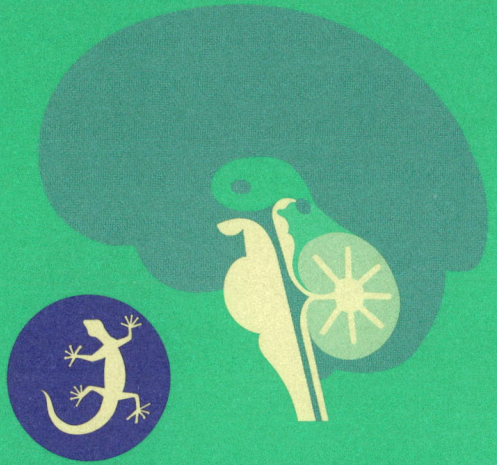

Is the left hemisphere logical and the right creative?

━━▶ **Both hemispheres do everything, but not usually as well as each other. The differences between them have sometimes been exaggerated, but there's some truth in the old idea about logic versus creativity.**

The two hemispheres of the brain are mirror images of each other – all the bits are present on both sides, and most of their physiological differences are microscopic. Over time, however, these slight differences lead them to specialise in different tasks.

In most people the left hemisphere, which controls the right side of the body, becomes dominant. It takes over the most obvious language tasks: speech, word comprehension and literacy, while the equivalent areas in the right become expert at detecting speech rhythm and emotional tone. The left tends to concentrate on details and plans; the right keeps a grasp on the wholeness of a situation. Negative emotions tend to arise in the right hemisphere and it tends to contribute a cautionary effect on the left hemisphere's observations, which are determinedly optimistic.

The left hemisphere's ownership of language suggests that it is responsible for conscious thought, and when it talks it expresses its own ideas, not those of the whole brain. In healthy people the two sides are kept in constant communication by swapping information through a central band of interconnecting axons called the corpus callosum, so they act as one.

A series of intriguing experiments using volunteers whose hemispheres had been surgically separated (for medical reasons, such as to control siezures) suggests there could be two personalities in our heads. The volunteers' unusual condition prevented information from passing from one side of their brain to the other, and the researchers, using complicated equipment, showed each hemisphere a different image. There was no internal communication between the hemispheres, and when a picture of, say, a fork was shown to their right hemisphere, the volunteers said they could not see anything. Yet when asked to identify the object by touch, using their left hand (controlled by the right hemisphere), they had no problem. The original researcher on the 'split brain' project, US neuroscientist Roger Sperry (1913–1994), remarked: 'Everything we have seen indicates that the surgery has left these people with two separate minds. That is, two separate spheres of consciousness.'

SPLIT BRAIN EXPERIMENTS

In his split brain experiments on people with surgically divided hemispheres, neuroscientist Roger Sperry displayed a different object to each hemisphere. The left, the seat of language, could name the object shown to it, but the right could not, because it was unable to pass the information to the left hemisphere to articulate it. The right hemisphere can, however, 'tell' by touch. When asked to select the object shown to the right hemisphere, using their left hand (controlled by the right hemisphere) the person grasped the correct object, even though they could not say how.

'WHAT ARE YOU HOLDING?'

SPEECH CENTRE

FORK

What are the main landmarks of the brain?

→ Some of the brain's important parts are obvious – the two hemispheres of the brain and the crinkly grey cortex. Others are hidden inside and can only be seen in cross-section.

The big round part of the brain is called the cerebrum, and the wrinkled skin is the cortex. A structure that looks like a small version of the main organ pokes out the back. This is the cerebellum. In front of it, the brainstem emerges to form the spinal cord.

The cerebrum is divided into two hemispheres by a deep valley that runs from front to back, called the longitudinal fissure, and each half has four lobes. These are separated by deep grooves known as sulci (singular: sulcus) and between them the cortex bulges into sausage-like structures called gyri (singular: gyrus). The sulci and gyri are used to locate functional areas of the brain. The fusiform gyrus, for instance, stretches along the lower edge of the left hemisphere and is known to process raw images into recognisable objects. The back part of the cerebrum deals mainly with sensory perception, and the front makes sense of it. Two adjacent strips curl over the cerebrum like Alice bands. The back one is the somatosensory cortex, which registers information coming in from the body, and the one in front is the motor cortex, which controls bodily actions.

Most of the important brain modules can be seen only in cross-section. Sliced down the longitudinal fissure, each side reveals, from top down, the inner fold of grey matter called the congulate cortex, and then a thick band of connective tissue, the corpus callosum, which binds the hemispheres and carries information from one to the other.

Under the limbic system there is a cluster of modules that between them keep us attuned and responsive to our environment. The thalamus takes eveything in and directs it for further processing, the hypothalamus controls our appetites and physical needs, the amygdala samples incoming information and produces appropriate emotional responses, and the hippocampus is the seat of learning and memory.

The brainstem, at the bottom, connects the brain to the spinal cord and controls basic survival mechanisms, including breathing, blood pressure and sleep/wake cycles.

BRAIN ANATOMY

The brain is a formidably complicated organ. It consists of hundreds of distinguishable nuclei (clusters of neurons) which vary greatly in shape and size. Most of them have internally intricate anatomy, parts of which may also be descried as nuclei. The physical constitution of the brain is still being discovered as microscopes and dissection techniques improve and show greater detail. Brain imaging technologies that allow us to look inside a living, working brain are slowly revealing what each part does.

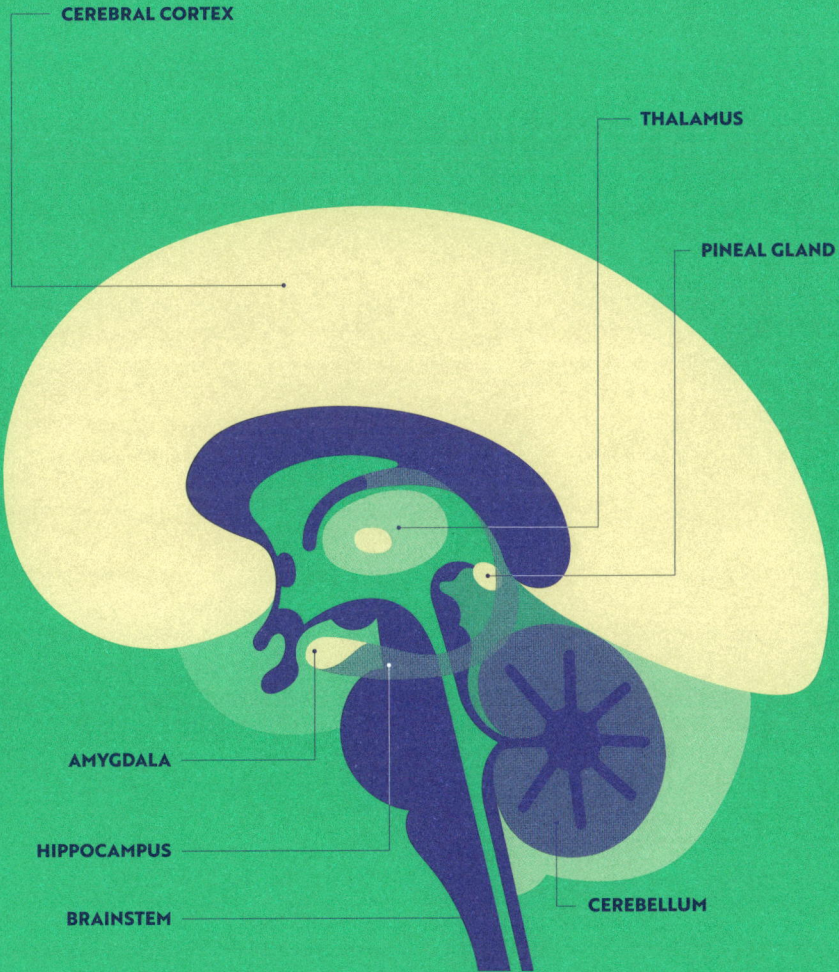

CEREBRAL CORTEX

THALAMUS

PINEAL GLAND

AMYGDALA

HIPPOCAMPUS

BRAINSTEM

CEREBELLUM

If my brain works on electricity, where is the battery?

→ The brain is commonly said to use 20 watts to function. Its 90 billion or so neurons are like tiny batteries, and the current they produce carries the signals that underlie all our experiences.

There are several different types of neuron – motor neurons that run down the spine, shorter ones that act as relays, sensory ones that bring information from the body. They do slightly different things and vary greatly in how they look, but their basic anatomy is similar.

All neurons have a cell body and fibres coming out of it. One of the fibres – the axon – carries electricity out to other neurons, and the other fibres – dendrites – receive it via receptors on their ends. Tiny gaps, known as synapses, separate the axons from the dendrites they connect with. Most neurons store chemicals called neurotransmitters in the axon terminals.

A neuron generates electricity when it 'fires'. It can do this dozens, even hundreds of times every second, yet the process is complex.

When they are at rest, neurons have a high concentration of negatively charged ions (molecules that have one or more extra electrons) inside their cell bodies, while the fluid around them has a high concentration of positive ions. If the cell is stimulated strongly enough, tiny channels in its outer membrane open suddenly and allow positive ions to flow in. This creates a tiny charge which flows to the tip of the axon, where it releases its neurotransmitters. The chemicals then carry the signal over the synapse and on to the next neuron, which may also fire. The chain of electricity may then continue along a neural pathway like a spark along a trail of dynamite.

Neurons that fire together are primed to fire together in future – a process called long-term potentiation. If the same neurons fire together repeatedly, their axons reach out to one another – physically stretching – and eventually form stable links. This is the basis of learning. Networks of neurons linked together through this process are the stuff of memories.

LONG-TERM POTENTIATION

When neurons repeatedly fire together, the connection between them is strengthened at the synapse, where they meet. No. 1 neuron (left) fires and sends neurochemicals over the synaptic gap to receptors on a dendrite of No. 2 cell, which fires in response. No. 2 cell increases receptors to pick up subsequent messages from No. 1 cell more easily. No. 1 releases more neurochemicals as it fires more frequently, and No. 2 has more receptors to pick them up. And so the 'conversation' flows ...

AXON   SYNAPSE   DENDRITE

Do we have a second brain in our guts?

⟶ No. But your gut is in constant two-way communication with your brain and has a profound effect on the way you feel. Sometimes it even knows something before you do.

The enteric nervous system – sometimes called the 'gut brain' – is a dense web of neurons embedded in the wall of the gastrointestinal system. It stretches from stomach to rectum and, although it can't do calculus or compose music, it affects your emotions, your health and your general well-being.

The main role of the ENS is to control digestion, from swallowing to excretion. It recognises nutrients and releases appropriate hormones and enzymes to absorb them, moves them through the digestive system by stimulating the gut to contract, and controls excretion.

The ENS also plays a major part in regulating the immune system, largely through its interplay with trillions of micro-organisms – bacteria, fungi, parasites and viruses – that live inside the gut. Some of the microbes are helpful and some are potentially harmful, and in a healthy gut the two are maintained in balance. If the balance is disturbed – perhaps due to infection, certain diets or antibiotics – we may feel mentally, as well as physically, unwell.

Although ENS functions are unconscious they are in constant two-way communication with the brain and influence many of our felt experiences. Butterflies in the stomach and weak knees at the sight of a loved one are produced as much by signals from the gut to the brain as the other way around. The fluttery feeling is caused by a drop in blood supply to the stomach as enteric nerves signal blood vessels to shunt more oxygen to the muscles in preparation for action.

The extent of neural communication between gut and brain is a fairly recent finding and it is thought to explain why people with irritable bowel syndrome are more likely than others to develop depression and anxiety. It might also explain why drugs such as antidepressants often bring relief to people with digestive problems, and why 'talking cures' such as cognitive behaviour therapy are sometimes successful in relieving functional bowel disorders.

GUT HEALTH

We need to maintain a balanced gut biome for physical and mental well-being. Some medications, notably antibiotics, can cause a relative shortage of good bacteria. Probiotics contain healthy bacteria and may help restore the balance, although studies suggest that most of the bacteria is destroyed in the stomach. A more effective method is to implant small amounts of faecal matter from healthy people directly into the gut via the anus.

Can our bodies stay alive without a brain?

⟶ Life-support machines can keep a person's body breathing and pumping blood around, but without a brain they would not be conscious. It would be life of a sort – but not one worth living.

You can take quite a lot of brain from a body without killing the rest of it. The parts that maintain crucial survival mechanisms are situated in the brainstem, and a person can survive severe damage to the rest of their brain yet continue to have 'vegetative' functions such as breathing, circulation and sleep/wake cycles. People in this condition – a persistent vegetative state (PVS) – usually die within six months, though some survive for years without medical intervention.

PVS is a very difficult thing for families and loved ones of the patient because it may seem to them as though the person is still alive, in the full sense of that word. This is because some of the brain's attention mechanisms are situated in the brainstem, and cause reflex actions that seem to indicate some inner life. The patient's eyes may follow you around the room, for example, or turn to look at the source of a loud noise.

PVS and the similar minimal consciousness state (MCS) allow a person's body to survive without most of their brain, but if their brainstem is severely damaged they will almost certainly die. In most countries the current legal definition of death is the complete and irreversible cessation of brain activity, which is diagnosed by testing for electrical activity in the brainstem.

It is possible for a body to be kept alive in the absence of brain activity by life support machinery. This has led to many tragic cases in which patients' loved ones have tried to prolong this 'artificial' life against the advice of doctors. In some cases the dispute has ended in court, where a judge has to rule on whether the machine should be turned off. When the machine is turned off, the body may continue to 'live' for an hour or so because the heart's intrinsic electrical system can keep it beating for a short while.

LIFE SUPPORT

Total brain death was established as the determining factor in most countries in the 1980s, but the age-old practice of checking for heartbeat and breathing is still the usual way to test for death. Life-support machines have saved many lives, but they have also introduced the problem of when it is ethical to 'pull the plug' on those whose bodies are permanently incapable of supporting them. A few people have been kept breathing on machines for years after officially being pronounced dead.

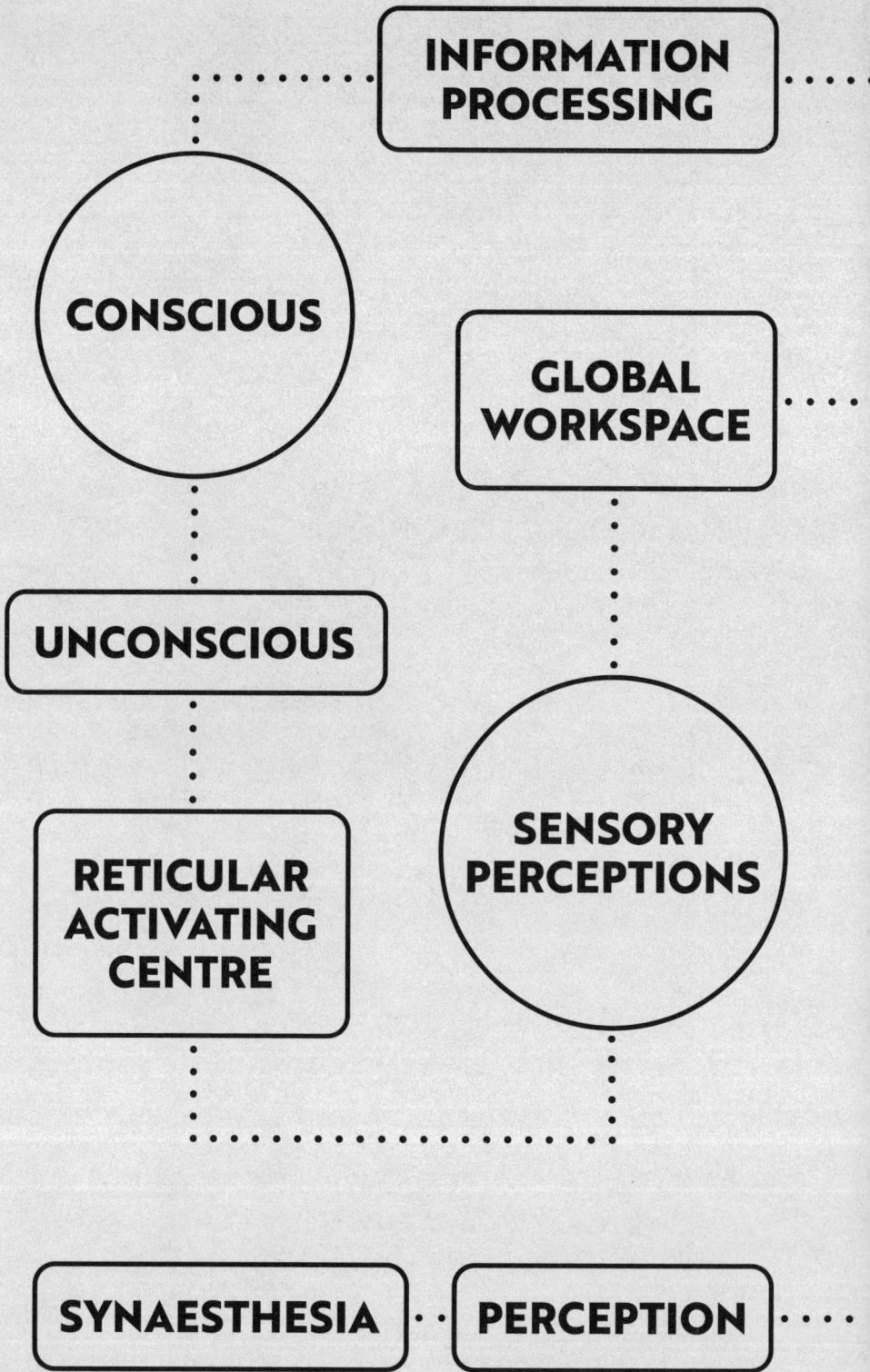

SURVIVAL

SENSES

VISION

CHAPTER 3

WORKING BRAIN

PROPRIOCEPTION

CONSCIOUS THOUGHT

INTRODUCTION

The world we experience is not the 'real' world. **SENSORY PERCEPTIONS** such as smells and tastes, sights and sounds are created – not detected – by the brain. They are the end products of the massively complicated **INFORMATION PROCESSING**, which ends in consciousness. This chapter describes some of what we know about it.

Our experiences are usually triggered when our bodies – particularly our sense organs (eyes, ears, nose, tongue and skin) – interact with the stuff of the universe. That stuff can only be fully described by mathematics, and until quite recently we didn't even know it existed.

Sound waves, for example, were not universally recognised until the 17th century, when the Anglo–Irish chemist Robert Boyle demonstrated that a ringing bell could be silenced by putting it in a sealed jar and pumping out the air. Until then many scientists thought the sound was created by its source and sent to the ear like a bullet from a gun. Molecules – the things that trigger the processes that create smells and tastes – were only proven to exist in the early 20th century, and why and how some of them, but not others, cause the brain to sense them as smells is still a matter of debate. Light waves were discovered around the same time, but the notion of the photon – the minute packet of electromagnetic radiation that activates the photoreceptors in the eye – was not known until Albert Einstein proposed it in 1905.

The interaction of our external sense organs and these physical forces are now well described, partly because these organs are relatively easy to examine. More recently we have discovered senses beyond the usual five; the subtle interplay between gravitational forces and our bodies that gives rise to **PROPRIOCEPTION**, for example, the so-called 'sixth sense'.

In this chapter we see how sense organs react to stimuli by generating streams of electricity which they send on to the interior of the brain, where it is much more difficult to see what is going on. Brain imagining techniques are starting to show what happens in this dark space, and in this chapter we chart some of that new knowledge.

To understand brain processes it helps to break them down, initially, into bits. Much recent neuroscience has therefore focused on the location of functions – which bits of the brain do what, from the massive dome of cortex to tiny hidden nuclei such as the amygdala. However, the brain works as a whole, like (and, indeed, with) the rest of the body. This chapter sketches some of the circuitry that allows this to happen. It also touches on the minute differences in brain anatomy and function that cause each brain to invent a very slightly different world.

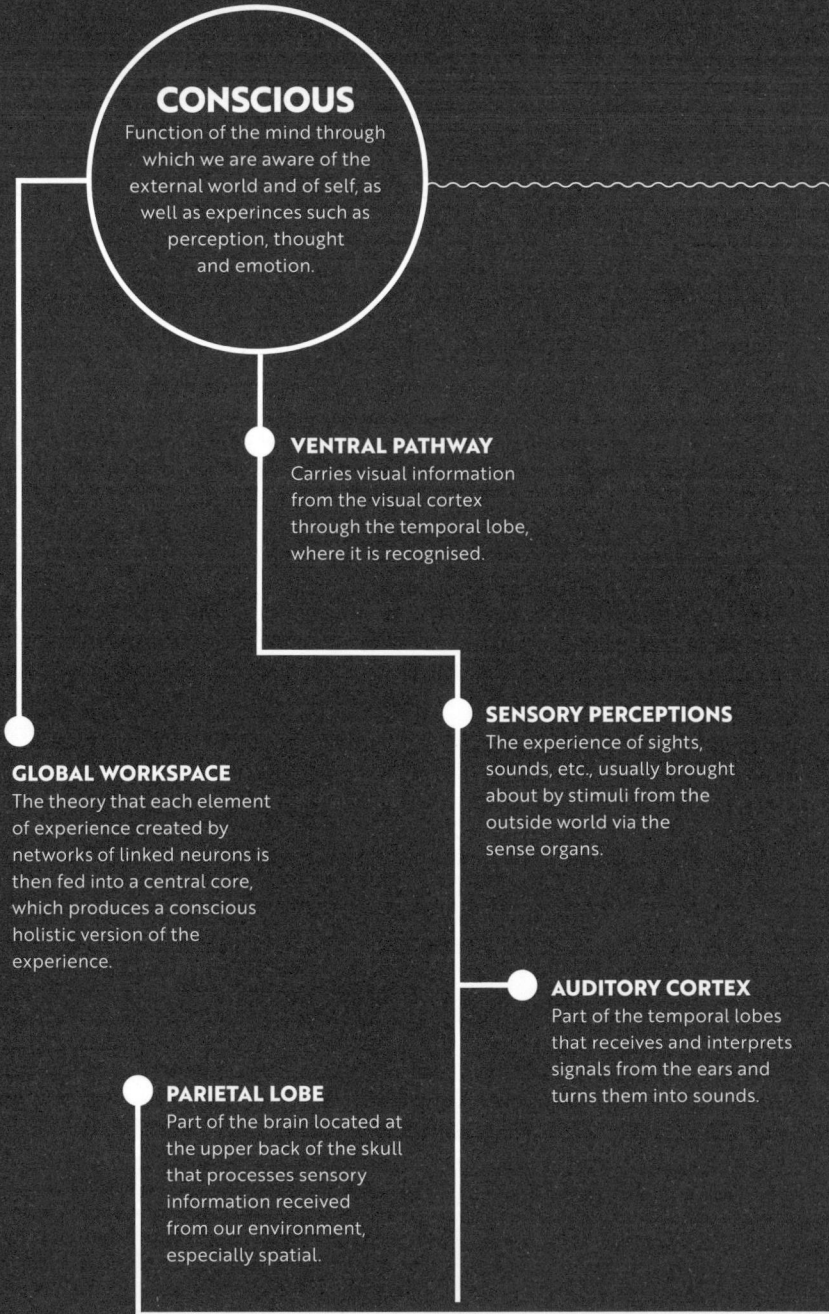

WORKING BRAIN MAP

CONSCIOUS
Function of the mind through which we are aware of the external world and of self, as well as experinces such as perception, thought and emotion.

VENTRAL PATHWAY
Carries visual information from the visual cortex through the temporal lobe, where it is recognised.

SENSORY PERCEPTIONS
The experience of sights, sounds, etc., usually brought about by stimuli from the outside world via the sense organs.

GLOBAL WORKSPACE
The theory that each element of experience created by networks of linked neurons is then fed into a central core, which produces a conscious holistic version of the experience.

AUDITORY CORTEX
Part of the temporal lobes that receives and interprets signals from the ears and turns them into sounds.

PARIETAL LOBE
Part of the brain located at the upper back of the skull that processes sensory information received from our environment, especially spatial.

EXTERNAL

EXPERIENCE

SENSES
The faculties by which we experience our outside and internal worlds, such as taste, touch, smell, sight and hearing.

DORSAL PATHWAY
Runs up and over the brain from the visual cortex, feeding visual and spatial information to the motor cortex to guide action.

SYNAESTHESIA
Crossover between two or more senses due to the connection of neural pathways serving different sensations; e.g., a tone may be experienced as a colour as well as a sound.

PROPRIOCEPTION
Generally agreed to be the 'sixth sense'; a bundle of senses that keep us aware of what is happening to our body, such as movement.

INTERNAL

UNCONSCIOUS
Brain activity of which we are unaware, even though it might drive behaviour; e.g., automatic movements such as walking while thinking of something else.

TEMPORAL LOBES
Middle-side part of the brain that processes auditory information, makes sense of language, stores semantic (words) memory and enables recognition of visual objects.

RETICULAR ACTIVATING CENTRE
Neural circuitry in the brainstem that controls alertness and the sleep/wake cycle.

OCCIPITAL LOBE
At the back of the brain, this lobe receives signals from the eyes and turns them into visual elements such as movement, colour and shape.

V1
Part of occipital (visual) lobe at the very back of the head where visual signals are initially registered before being sent on for processing by other brain areas.

What does the brain actually do?

→ Our brains exist primarily to keep our bodies alive. In most animals their activity is directed almost entirely to that. Human brains have become more complicated, however, and can do a whole host of things not directly concerned with survival.

The primary business of the brain is to receive information about what is happening and adapt our bodies to it. Sometimes it sends out instructions to the muscles to move. Between this 'information in' and 'action out', it processes multiple streams of information in parallel.

Information about the outside world comes via our sense organs, while knowledge of what is happening in our inner world comes from a complex system of peripheral nerves and chemical messengers. We usually pay attention to our inside only when the information from it is pressing, yet this internal monitoring is the most essential thing the brain does. It guards us against physical harm by generating pain, keeps our cells fed and hydrated by making us hungry or thirsty, and helps our bodies stay upright when we stand, balanced when we walk and co-ordinated when we move.

While all this is happening, the brain processes a deluge of signals coming in from outside. One of the first things it does is to 'taste' incoming information to see if it suggests a threat or an opportunity, such as food or a sexual encounter. If one of these is detected it sends signals to the body to prepare for it. A threat may trigger a physical reaction like running away or hitting out, for example.

As well as the moment-by-moment apprehension of the world, our brains also encode our experiences as they happen, forming memories. If a new experience chimes with a previous one, it uses the memory to decide what to do. If a memory is regularly rehearsed it is learned – either deliberately, by repetition or, if the experience is very intense, automatically.

The brain does all of this, all the time, mostly unheeded. Conscious feelings, perceptions and thoughts are extras which, in humans, do not always help the brain to fulfill its primary purpose, which is to keep us alive.

THE UNCONSCIOUS

The vast majority of congition is unconscious. We know nothing about the enormous, continuous process of translating streams of electrical information into a representation of the world that allows us to survive in it. Indeed, most of what we do is done entirely without our knowledge: for example, the hundreds of movements your brain directs your body to do in order to take a step forward; the intricate calculations in your cochlear and temporal lobe in order to hear a sound.

CONSCIOUS

UNCONSCIOUS

How does the brain tell us what's outside?

➔ The outside world is full of light waves and sound waves, molecules that impact our tongues and noses, and objects and forces that interact with our bodies. The brain turns these physical phenomena into sights, sounds and smells.

The world we move through is virtual reality, not what is really there. The brain constructs the environment moment by moment, starting deep in its basement, then flowing into the spinal cord, where information enters the brain from the body.

Mini-organs in the brainstem regulate our basic survival mechanisms – breathing, heartbeat and sleep/wake cycles and, crucially, our attention. A cluster of cells called the reticular activating centre (RAS) continuously signal other parts of the brain to orient our senses to things that might matter, such as a sudden movement or a loud noise.

Once our attention is directed to some part of our world, our sense organs – eyes, ears, nerve endings, tongue and nose – pick up information from it and turn it into electrical signals which, with the exception of those coming from the nose, are passed to the thalamus. The thalamus sits in the limbic system and acts as a central guidance unit. It relays the information it receives to specialist areas of the cortex where they are processed further. Signals from the eyes are sent to the occipital lobe in the back of the brain, those from the ears go to the auditory cortex, on the side, and messages from the body are sent to the somatosensory cortex – a strip of tissue that wraps aound the brain like an Alice band. Information from the nose bypassses the thalamus and goes directly to the limbic system.

Some information from our sense organs becomes conscious sensations, but most is used unconsciously, to adapt our bodies to our environment. Information about temperature, for example, causes capillaries to expand or contract, and light signals change the size of our pupils.

Neurons transport electrical signals at speeds of up to 200 metres (650 feet) per second, but it still takes nearly half a second for the brain to transform raw information from our sense organs into conscious perceptions.

THE SENSES

Our sensory organs – eyes, ears, nose, tongue and the nerve ends in our body – are constantly impacted by physical forces such as light and sound waves. They begin the massive task of constructing a sensational world by generating electrical signals which are transmitted to the brain. The signals are distributed to areas that are expert in interpretation, to inform us about what is 'out there' in the world.

How do the different parts of the brain work together?

→ Brains contain hundreds of modules, each of which has specific functions. They do not work independently, though. Cognition is like a musical symphony – every part of the orchestra contributes.

Each part of the brain specialises in specific things. The hippocampus, for example, encodes experiences and retrieves memories, and the amygdala is the alarm that kicks off emotion.

On their own, though, no single part does anything. Every experience, however momentary, is constructed by multiple brain areas, and before an experience becomes conscious, the elements each part produces – sensory perceptions, emotions and thoughts – are bound together to make a meaningful whole. This bonding is essential for us to make sense of the world. If it didn't happen, we would experience everything as disjointed bits. If a red car backfired in the vicinity, for instance, you would be aware of a patchwork of shape, colour and noise but have no idea what it was.

One of the essential things needed for experiences to be comprehensible is that the neurons that produce each of its elements are firing at the same time and at the same rate. To construct the noisy red car, for example, colour neurons would fire at the same time and speed as neurons in the shape area in the occipital lobe, neurons in the location area (parietal lobe) and neurons in the auditory cortex (temporal lobes). The car would probably stimulate activity in the amygdala too.

Synchronous activity is not enough to produce a conscious experience, however. Current research suggests that for an event to be known about – not just acknowledged unconsciously – a fairly large, linked group of neurons must be firing to produce each component. These, in turn, are pulled together in association areas. One idea is that they are all pulled together into a core network of cells, referred to as the 'global workspace'. It is rather like a hierarchical workforce: groups of underlings each produce a report on one aspect of a project and together pass it on to a small team of managers who condense it into a single report. Their report is then sent out to the rest of the brain as though for consultation, and when everyone has had their say, it becomes an experience.

FUTURE   PRESENT

**GLOBAL
WORKSPACE**

VALUE   PAST

FOCUS

THE GLOBAL
WORKSPACE

*Each element of experience is
created by networks of linked
neurons, firing together. These are
then fed into denser core networks.
One theory is that there is a central
core, called the global workspace,
which produces a holistic version
of the experience in which all the
elements are bound together. Other
theories hold that active neurons
constantly communicate between
themselves so that each experience
is like a chorus of voices that
together produce a coherent whole.*

Do we have a sixth sense?

⟶ If you mean the ability to see supernatural goings-on, almost certainly not. If you mean a sense in addition to the usual five (seeing, hearing, tasting, smelling and touching), then yes!

Senses are traditionally understood as the information we take in through our visible sense 'organs', hence the 'five': light via eyes; sound thanks to ears; taste by the tongue; smell by the nose; and bodily sensation through the skin. We now know we have many more senses than that, delivered by less obvious means.

Some psychologists have identified more than 50 human senses, but that extends the definition of sense to include things we feel after information has been at least partially processed in the brain. An example would be sensitivity to seasons. Although we don't consciously work out what time of year it is, the brain combines several different sensory streams to tell us. A narrower definition of sense – raw knowledge that affects the brain directly – suggests there are a dozen or so.

Proprioception is generally agreed to be the 'sixth sense', though it is more accurately a bundle of senses. Together they keep us continuously aware of what is happening to our body. Most of this is done unconsciously, but we notice it immediately if it fails, rather as we notice the hum of a fridge when it suddenly stops. If the signals relating to balance (equilibrioception) stop, we fall over, and if our sense of movement (kinesthesia) is interrupted, we don't know we are moving.

Similarly, specialised receptors in our skin tell us about external changes. As well as simple touch, they send signals about temperature (thermoception), potentially damaging stimuli (nociception, felt as pain), pressure, irritation, tension and stretch. Once sensory information is received by the brain, it is combined with other information to provide an account of what is going on.

Everything we experience is rooted in sensation. Sometimes a sensory stimulus is experienced directly – when you are dazzled by light, for example, or prick your finger – but most of the time our brain instantly integrates sensory input with our thoughts, memories and emotions to give us a multi-layered whole experience.

PROPRIOCEPTION

Proprioception is sometimes called our sixth sense. It is rarely experienced consciously; most of the time it hums along in the background, keeping our bodies functioning by making thousands of miniscule adaptions to position and balance every minute. We usually only get to know about it when it fails. Certain drugs, notably ketamine, the anaesthetic widely used as a leisure drug, can cut off the feedback from body to brain, leading to a terrifying sensation of 'lost body'.

How do we make sense of the signals that enter the brain?

⟶ The brain processes the information it receives to guide our actions, but it doesn't have to 'make sense' of it in order to do that.

The world we know is made of recognisable parts – visible objects, sounds and bodily sensations. Most of the information we take in, however, is registered, used and discarded before it becomes such a part – it never gets beyond being a raw stream of electrical signals.

Take a flicker of light reflected off a dust particle that is moving towards your eye. The light enters your eye, where retinal cells turn it into electrical signals which zoom along your optic nerve into the inner sanctum of the brain. Very quickly the signals arrive at the thalamus, which, like a police officer on traffic-control duty, directs most of them towards the vision centre right at the back of your brain. Some signals, however, get shunted off to minor pathways that go to brain areas which trigger the muscles around your eye to make you blink. Job done, the signals fade away before you are aware of

what happened. If the light had bounced off a speeding ball, rather than a dust particle, it would have stimulated much stronger signals, which would direct muscles all over your body to produce a flinch.

Note: There is no 'dust' or 'ball' involved in this; no object that you would recognise. The eyeblink and the flinch are kicked off entirely by automatic mechanisms triggered by raw light-borne signals. Furthermore, it happens before you know consciously that there is even a ball in the vicinity.

Meanwhile, if the signals are strong (as they would be if the light came from a ball), most of them continue to the occipital, or visual cortex, at the back of your brain. When they hit the furthermost area, called V1 ('V' for vision), they are amplified and radiated to a wide range of brain areas. There is still no ball, but V1 starts the process of constructing one.

UNCONSCIOUS SIGHT

Unconscious sight – like that which makes you blink – uses an evolutionary ancient neural pathway that carries light signals direct to areas of the brain before they go to the visual cortex, which produces conscious vision. Some people with no conscious sight, due to brain injury, have learned to 'see' in this way, to the extent they can recognise shapes and even read simple words.

How does the brain turn sensations into thoughts?

→ **Very few sensations are experienced as conscious thoughts – the vast majority of sensations registered by the brain are junked before they become conscious. Those that make it through are usually couched in language.**

A yellow tennis ball is speeding towards you. Your body is already flinching because some of the electrical signals carrying this information went straight to areas of your brain that trigger reflexive movement. Now the remaining signals arrive at V1, the very hindmost bit of your brain, which broadcasts them to the rest of the brain.

Two major pathways carry the news: the dorsal pathway goes up and over the brain to the motor cortex, while the ventral pathway heads down and along the bottom edge. The dorsal path ends in a conscious movement – hitting the ball, for example. The lower path produces the thought: 'Yellow ball approaching.'

Before that thought can occur, the signals on the ventral path pass through a stretch known as the recognition area. Here the encoded information – telling of approaching yellow round things – resonates with similar encoded information that's been stored after previous experiences. Memories of balls, and the consequence of being hit by one, may be among them, or at least there will be something close enough to start forming an idea of what's happening. Along with the memory comes emotion – fear or maybe anticipation – and heightened alertness.

The next area of brain to get excited by the ball is concerned with language. Here the yellow roundness resonates with the encoded words 'tennis ball', and maybe a few others. Now you have all the ingredients for a conscious thought. The brain still has a lot of work to do before you think it, however.

The sensory elements of a perception, along with the associated emotions, memories and words, come together in the brain's frontal lobes, along with information from the dorsal pathway about location, motion and relevance to the body. Now the thought emerges: 'Yellow tennis ball approaching.' It has taken nearly half a second.

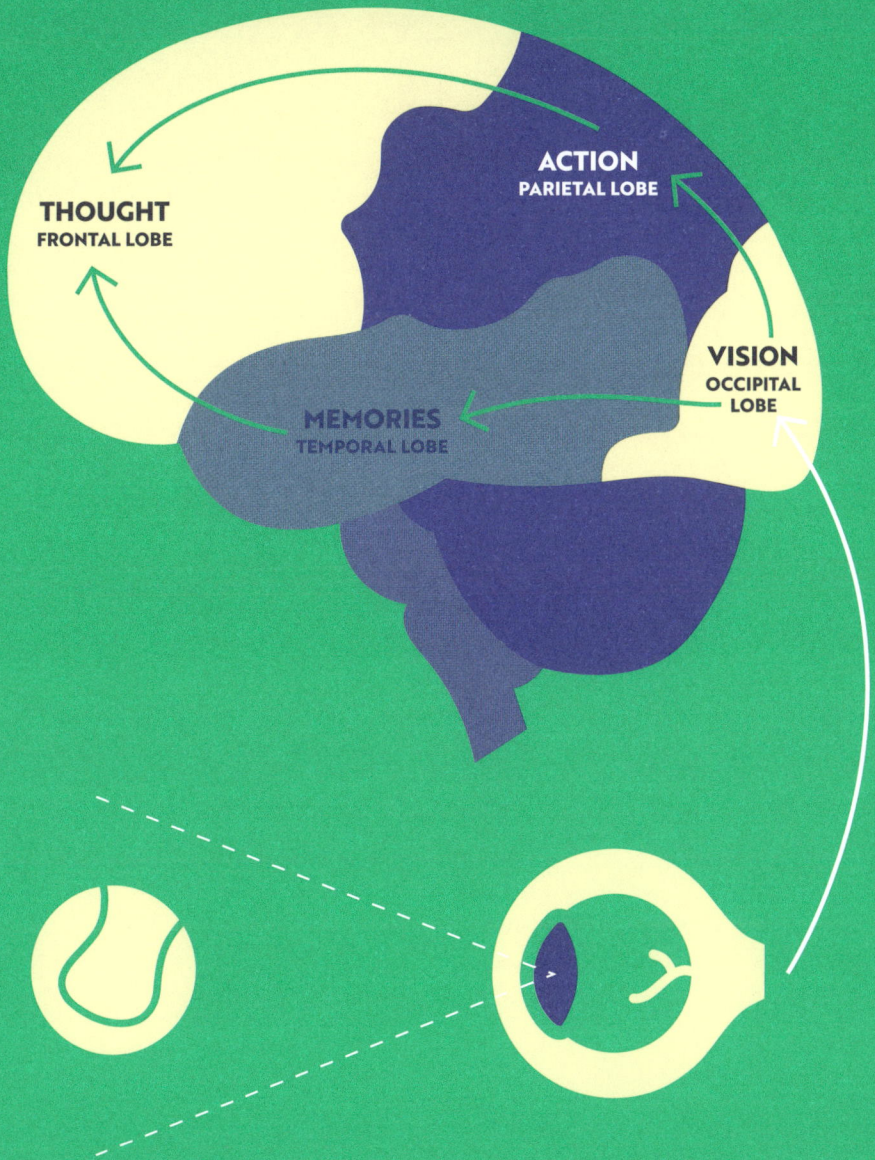

THOUGHT
FRONTAL LOBE

ACTION
PARIETAL LOBE

VISION
OCCIPITAL LOBE

MEMORIES
TEMPORAL LOBE

CONSCIOUS THOUGHT

Light-borne information, such as a fast-approaching tennis ball, enters the eyes then travels to the occipital lobe at the back of the brain, which turns it into visual elements such as movement, colour, shape and position. When the perception is constructed, it is passed through areas devoted to memory, in order that it is recognised, then to the frontal lobes, where it may trigger a thought. In the case of the tennis ball, the thought may be 'Hit it!'

Do we all see the same world?

→ No two brains are identical, so no two perceptual worlds are identical. Most of us perceive things similarly enough to live in a shared environment without confusion, but it's likely that everyone senses the world slightly differently.

My brain might respond very strongly to colour, for example, while yours may be particularly sensitive to movement. Our differences may be present at birth thanks to our varying genetic make-up. Some gene mutations make certain foods taste bitter, while others affect the intensity of perceived pain. Low-level cognitive distinctions may lead to different high-level preferences. My strong response to colour might lead me to love artwork, while you may be thrilled by dance.

Conversely, high-level cognition – such as tastes in art – can affect basic perception by remoulding the brain. If you watch a lot of ballet, you will prime the brain areas that process movement to become more densely connected, and as a result you will see tiny movements that would be lost on someone else. Similarly, people who work with moving images can often spot whether a film is shot using a digital or a film camera, while a casual watcher is unlikely to see the difference.

Usually, we only come to know about individual differences in perception when a person experiences the world in a notably unusual way. For example, some people 'see' sounds or 'taste' words. This strange condition is called synaesthesia, and it occurs when areas of the brain concerned with different senses are stimulated by one another. One theory is that in people with synaesthesia different sensory areas are connected by neural pathways that, in most people, get pruned away in infancy.

To some extent, we can never know for sure how our own perceptions differ from those of others. We depend most of the time on words, but they are usually too clunky to capture accurately our fine-grained perceptions. Is your 'red' my 'red'? We'll never know for sure.

SYNAESTHESIA

When a person has a crossover between two senses, it is known as synaesthesia. For example, they may see a particular colour when they hear a particular word or taste a specific thing when they touch a certain texture. It is known to affect one in fifty people, although someone might be synaesthetic and not know it – their multimodal world is just 'how things are'. The condition occurs when the neural pathways serving two different sensations are connected. It might be a hangover from infancy, when the brain was densely interconnected.

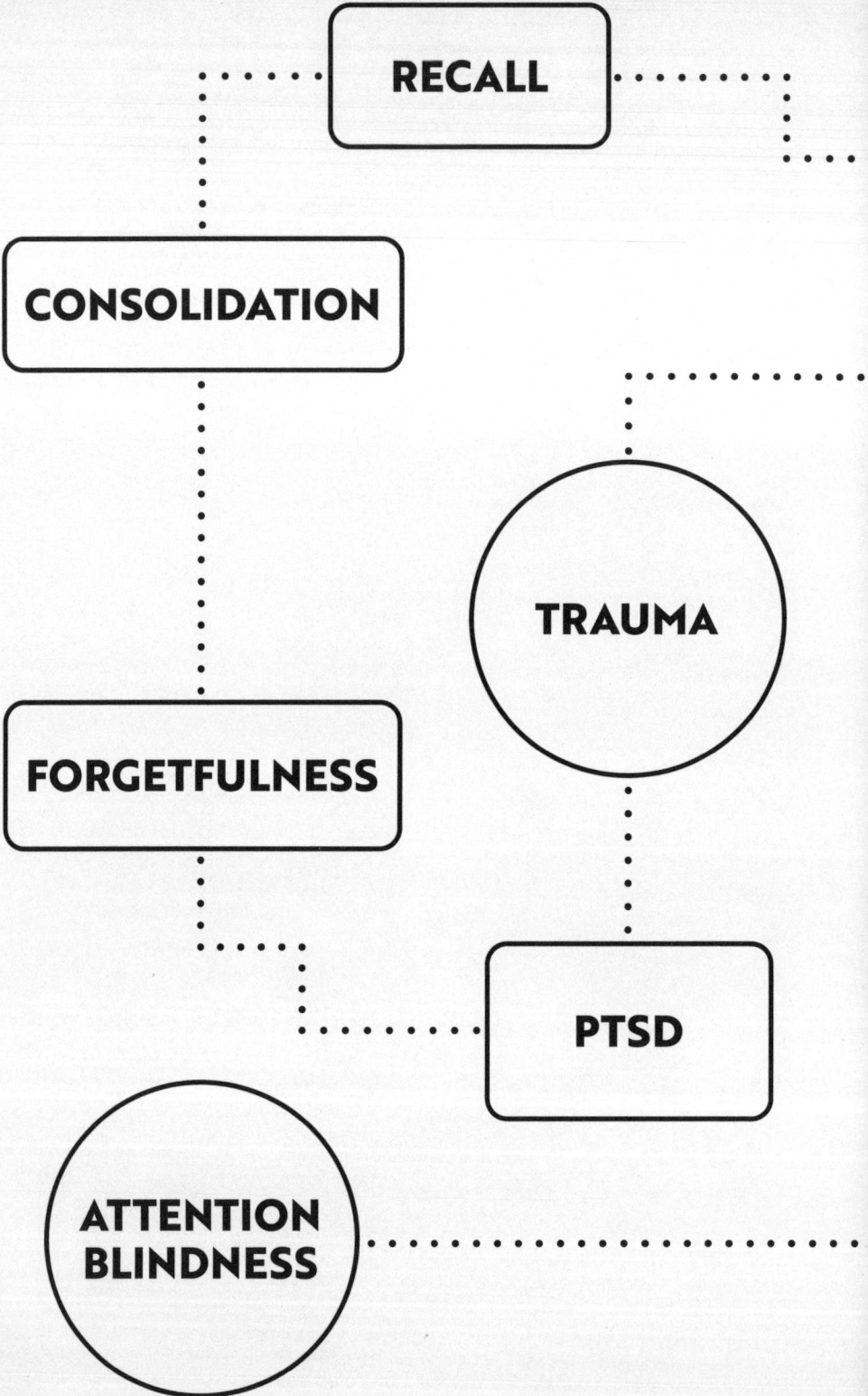

CHAPTER 4

MEMORY & LEARNING

MEMORY

MEMORY DISTORTION

CAPGRAS SYNDROME

PROSOPAGNOSIA

FUSIFORM FACE AREA

INTRODUCTION

We have always looked to metaphors to help us understand new discoveries: by likening something we don't understand to something we do, we get some kind of handle on it. **MEMORY** has been likened to almost every method we have ever devised for storing information, from wax tablets, filing cabinets and libraries to quantum computers. One of the most enduring comparisons is with video recorders. The machine's laying down of events and subsequent playback seems to match our experience. Subjectively it seems that we 'record' information and experience, store it in our brains as though on a tape, and replay it when we recollect it. Most of the time our memories are kept out of mind, like a videotape shoved in the back of a dusty cupboard.

As with so many things about the human brain, however, our subjective impression of memory does not match the objective facts. For one thing, we do not remember everything – in fact we remember very little, and what we do remember is mainly then forgotten almost instantly. Some experiences are physically encoded in the brain as on a tape, but they are not stored in neat, sealed packets. Rather, they are distributed around the brain and even shunted from place to place.

The main way in which biological memory differs from mechanical recordings is in recollection. We hardly ever remember an event accurately. Every time we think about something from the past it is tainted by the context in which we remember it. The mood we are in and the things that may be happening around us get wrapped into the memory and when it falls back into unconsciousness, those changes are locked into it, so that next time we **RECALL** it we experience it differently.

The plasticity of human memory is not a defect – it has evolved to help us adapt to an ever-changing world, and in this chapter we look at how and why it works that way. We explain why the brain keeps some memories out of our consciousness until a trigger brings them rushing back – so-called **RECOVERED MEMORIES**. It also describes how the brain can **DISTORT** or even fabricate memories without the person having any idea that they are false.

The complexity of human memory and learning means that, inevitably, it is subject to glitches. The most common complaint is **FORGETFULNESS**, and here we see why some people are more likely to forget things than others, and how very particular types of forgetfulness – the inability to remember faces, for example, as is the case with **PROSOPAGNOSIA** – may occur in otherwise healthy brains.

MEMORY & LEARNING MAP

ENCODING

MEMORY
Experience that is encoded in the brain and may be retrieved or re-experienced subsequently. Short-term memories are encoded as persistence of electrical activity; long-term as physical changes.

EPISODIC MEMORY
Personal experiences encoded in the brain and re-experienced, in whole or part, at a later date.

RECALL
Retrieval of memory, which can be triggered or prompted.

WORKING MEMORY
Very short-term; involves holding something in mind by prolonging a pattern of electrical activity for just as long as necessary.

PROCEDURAL MEMORY
Learned sequences of body movements such as riding a bike; once mastered they are encoded, usually permanently, in brain areas concerned with movement.

CONSOLIDATION
Process by which information is embedded in long-term memory that involves integrating new matter with existing stored information in the cortex and can take up to three years.

SEMANTIC MEMORY
Long-term memory for factual information; what's left of episodic memories that have lost sensory and emotional content.

IMPLICIT MEMORY
Long-term, unconscious and automatic; the things we don't know that we know but that can influence our thoughts and behaviour.

RECALL

FORGETFULNESS

Occurs when a memory fades due to degeneration of the neural links that encode it or becomes irretrievable or changed beyond recognition. Excessive forgetfulness may be due to brain damage or disease.

POST-TRAUMATIC STRESS DISORDER

Anxiety disorder that may involve a spontaneous re-experience of a traumatic event, the memory of which is stored in the amygdala and cannot usually be consciously retrieved.

ATTENTION BLINDNESS

Failure to be aware consciously of events outside one's focus of attention. A natural phenomenon that always accompanies concentrated cognition.

FUSIFORM FACE AREA (FSA)

Patch of cortex that lies along the ventral processing pathway where objects are recognised; amplifies information from faces to extract maximum information and emotional content.

PROSOPAGNOSIA

Impairment of face recognition, usually due to abnormal lack of activity in the FSA in the ventral pathway. Faces may not be recognised or may appear distorted.

CAPGRAS SYNDROME

Delusion in which a person believes others to be imposters or aliens, even though they acknowledge that the 'imposter' is visually identical to someone they know.

RECOVERED MEMORIES

Previously unavailable to the conscious mind due to having being repressed or forgotten but may be recollected when triggered by something similar or deliberately sought out, often in therapy.

MEMORY DISTORTION

As memories are reconstructed, they are vulnerable to being altered from the reality by factors such as misinformation, hindsight and suggestion. All memories change over time as they are integrated with new information.

ERROR

What is memory?

→ Memories are traces of experience that are physically encoded in the brain, and learning is the process of creating them. Unlike a recording device, the traces may disappear, change or become irretrievable.

Most of our experiences are forgotten almost as soon as they happen. A few, however, become memories. It is important to distinguish between having a memory and recalling a memory.

There are many different types of memory – the phone number we hold in mind just long enough to dial it, the distant recollection of a childhood tumble, the ability to ride a bike and the words we use to identify things. Each type uses different mechanisms and is stored in different parts of the brain, but what they all have in common is their form: memories are particular patterns of neural activity that are primed to occur. Remembering is what happens when that pattern of neural activity *actually* occurs. So, you can have a memory even when you are not recalling it.

Take a memory for a particular event. The initial experience caused simultaneous activity in a vast number of neurons throughout the brain. When neurons fire together they undergo molecular changes that make them more likely to fire together again. This tendency for a group of neurons to fire together is a memory. If the firing pattern is not repeated for a while, the molecular changes revert and the tendency for those neurons to fire together – the memory – is lost. This is forgetting.

Most events are forgotten, but if a firing pattern is repeated very frequently or if it is very strong, the neurons involved extend their axons towards one another and create physical links. These make it easy for electricity to flow between them, and once such neural pathways are established, memories can be retained for a lifetime.

Although many neurons involved in experiencing and replaying an event are in the cortex, memory formation and retrieval is largely dependent on the hippocampus in the limbic system. In memory formation it functions like a recording device, taking note of the firing pattern as though scoring a piece of music. In recollection it behaves like an orchestra conductor directing the instrumentalists to play the music.

MEMORY FORMATION

Each experience is a precise pattern of activity in millions of neurons. Most patterns persist just for the moment (sensory register) before fading away (forgetting). Short-term memories persist for a while, sometimes just long enough to act on the experience (holding a phone number long enough to dial it). A few patterns are encoded by the hippocampus, much as a listener would make a note of a piece of music so it could be played again later. If it is played repeatedly, the code is transferred to the cortex and becomes a long-term memory.

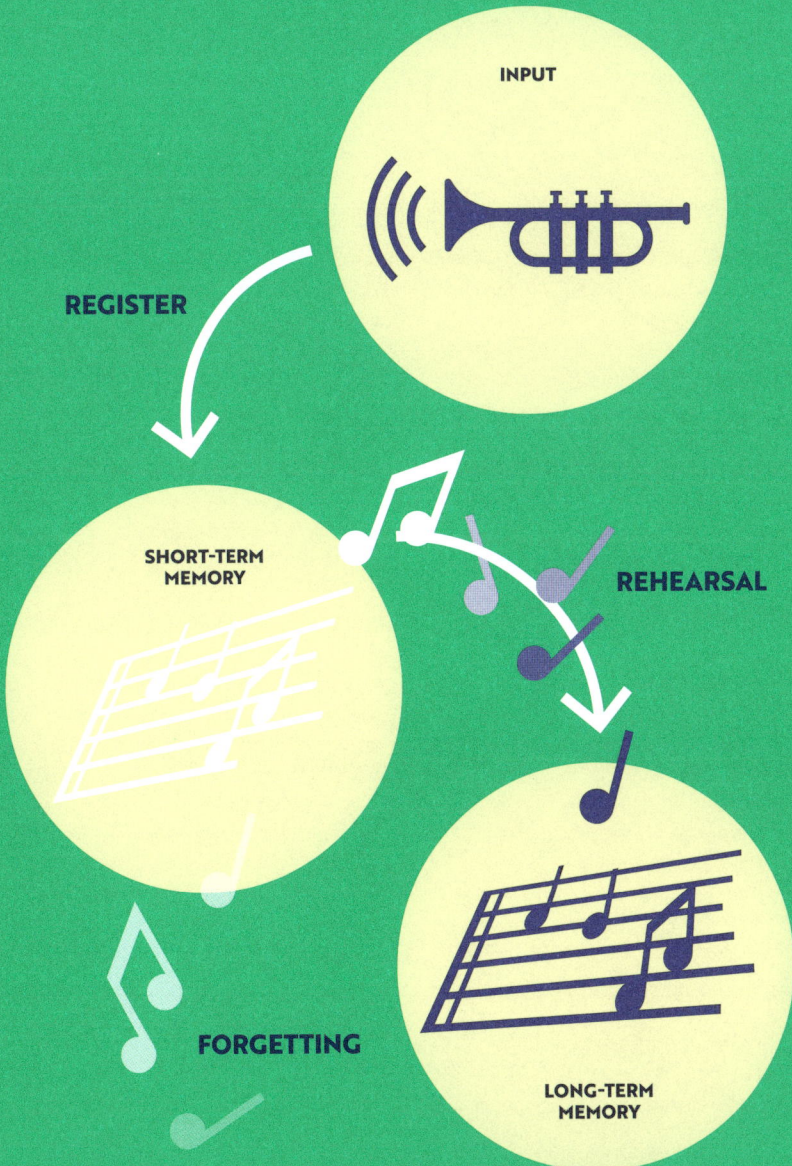

INPUT

REGISTER

SHORT-TERM
MEMORY

REHEARSAL

FORGETTING

LONG-TERM
MEMORY

Are there different kinds of memory?

→ Yes, there are. Working memory fades once the information is no longer needed; procedural memories stick for a long time, but all memories may be held for short or long periods, or even permanently.

Working memory is very short-term. It involves holding something in mind for just as long as necessary. An instruction to 'turn left at the crossroads' is a working memory, which will be replaced when you start following another instruction, such as 'continue straight on for a mile'.

A few working memories trigger the formation of less transient memories. If on approaching the crossroads a second time you remember turning left before, it is because the first experience created an episodic memory. Episodic memories are potential replays of your personal experience of past events. The crossroads example will contain not just the fact of turning left but the experience of it – how it looked, how it felt and what you thought as you did it.

Episodic memories are encoded initially by the hippocampus, but the hippocampus keeps re-stimulating the cortical neurons that produce experiences, and can do this to the point when the cells no longer need prompting. This process, consolidation, can take up to three years. Once it is complete, a memory is retained indefinitely.

Some memories become consolidated as 'semantic' knowledge. If, for example, you frequently repeat the journey via the crossroads, you will eventually 'know' to turn left without recalling previous experiences of doing it. This – the knowledge alone – is a semantic memory. It is what's left of episodic memories that have lost sensory and emotional content. Every word and fact you know was once embedded in the episodic memory of learning it, but unless a learning experience is very emotional, all that usually remains is unadorned knowledge.

Most working, episodic and semantic memories are 'declarative' or 'explicit', meaning you can describe or name them. Another type of memory is called 'procedural' and can't be conveyed in words because procedural memories are learned sequences of body movements such as riding a bike or playing drums. Once mastered they are encoded, usually permanently, in brain areas concerned with movement.

You may not consciously remember a memory but find that it 'comes back' when needed, and some memories remain unconscious but 'prime' our behaviour. For example, at least one case has been reported of a person with apparent total amnesia who was able to dial their home number when offered a phone, even though they had no conscious memory of it.

TYPES OF MEMORY

Human memories can be divided into different categories, although the physiological processes involved in forming and recalling them are the same. Short-term memories disappear before they are encoded. Long-term memories may be explicit, meaning they can be recalled at will; or implicit, meaning that their recall is not in words but in actions. A phobia, for example, may be an implicit memory of something that may not be consciously recalled but can still produce fear if activated.

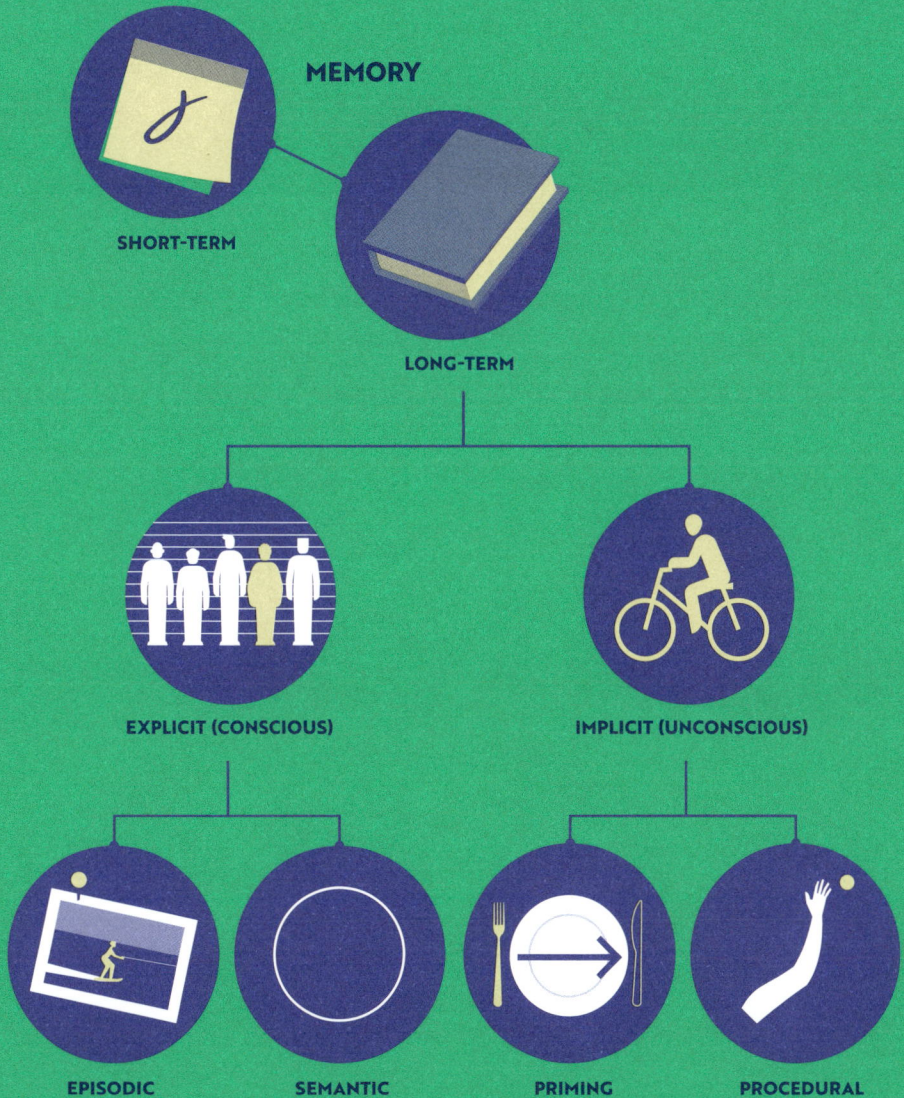

MEMORY

SHORT-TERM

LONG-TERM

EXPLICIT (CONSCIOUS)

IMPLICIT (UNCONSCIOUS)

EPISODIC

SEMANTIC

PRIMING

PROCEDURAL

Why are some people forgetful?

→ There are many different types of forgetfulness, and many ways to do it. The memory might never have been made, or it has faded, become irretrievable or changed beyond recognition.

Forgetfulness may be due to brain damage – stroke, trauma or dementia – that destroys the tissue in which memories are held or the pathways needed to retrieve them. It's not always a sign of illness, though – some forgetfulness is normal, and probably necessary to free up space.

To memorise something you must first encode it, and that doesn't happen automatically. Something extraordinary can happen in front of your eyes and if you are concentrating on something else, you might not notice it, let alone remember it. This is called 'attention blindness'. A famous demonstration of this was contrived in 2004 by researchers Daniel Simons of the University of Illinois Urbana-Champaign and Christopher Chabris of Harvard University, who staged a ball game and instructed an audience to count the number of passes made by one of the two teams. As the game progressed, a person in a gorilla suit walks slowly in front of the players, beats its chest, bows to the camera and walks away. Most people, when they first watch the experiment, fail to see the gorilla at all.

'Attention blindness' affects us all the time – we only see what we attend to. Our memories are inevitably selective too.

Those experiences that are attended to are noted and encoded by the hippocampus, in the brain's limbic system. It constructs memories by amplifying and prolonging neural activity patterns – the sort that occur when we are attending to something or repeating something deliberately in order to learn it. For the first few weeks or months, memories are stored in the hippocampus, and during this time they are easily forgotten. Over time, episodic and semantic memories are transferred from the hippocampus to permanent storage in the cortex, after which they are there permanently unless the cortical area is damaged.

THE CASE OF 'H.M.'

New memories cannot be made without the hippocampus. Before this was known, a small number of people had this part of their brain removed during surgery to control catastrophic epilepsy. The most famous case is that of Henry Gustav Molaison, known as H.M., who had his hippocampus removed in 1953, when he was a young man, to curb epilepsy. Afterwards, he could remember events from his childhood but was never able to form a new memory. He died in 2008, still thinking he was in his twenties.

Why can't I remember faces?

→ There's a particular part of the brain given over to recognising people's faces, so if you can't remember them at all, it may be because that part is not working properly.

Human faces are especially interesting to our brains, so much so that when one comes into view, it is subjected to close scrutiny by a part of the brain that has become known as the fusiform face area (FFA). This patch of cortex lies along the ventral processing pathway that carries visual information from the occipital lobe towards the frontal lobes. The FFA not only extracts more information from faces than from other objects, it also brings in emotional content, so that faces, more than other objects, are freighted with value: fear if it's frowning; attraction if it's smiling; and if it's known, familiarity. Usually a familiar face is subsequently shunted on through the language areas, where it is labelled with its name and/or role: friend, father, postman.

Prosopagnosia is the inability to recognise faces. It happens when something isn't working normally along the ventral pathway. A problem during early processing, before the information gets to the FFA, may cause faces to look weird or distorted, and the person may only know they are seeing one by consciously looking for a roundish object with a particular pattern or features. A problem in later stages of processing, when the signals are further along the recognition pathways, may allow a person to recognise that the face is that of a woman rather than a man, their age, and they may even have a 'gut feeling' about whether they loathe or love them – but they won't know who it is. People with serious prosopagnosia cannot recognise their nearest and dearest, let alone friends and acquaintances.

Capgras syndrome is a particularly weird form of face-recognition dysfunction in which the person recognises a known face but the face is not attached to a sense of familiarity. The people they know seem alien and almost 'not there'. In one tragic case, a sufferer murdered his father, convinced that he was a replicant.

PROSOPAGNOSIA

People with severe prosopagnosia may walk right past their nearest and dearest without knowing, as they are unable to recognise faces. Usually, people with the condition develop complex strategies to get around the social problems their condition causes. For example, they might memorise the clothes or a particular bodily feature in order to distinguish familiar people. One man, for example, always insisted his wife wore a yellow ribbon in her hair before they went to social gatherings, after once taking the wrong woman home from a party.

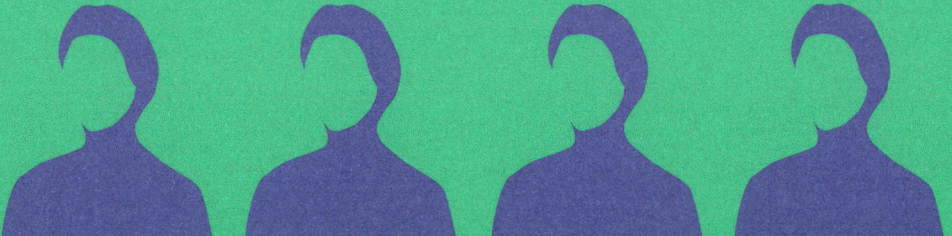

Do we store everything we know for ever?

→ No, most things are forgotten. We only store things permanently if they go through a process of consolidation, which takes a long time – up to three years.

The brain's ability to store information is vast. Long-term memories are held in it like cobwebs – strands of neural fibres connecting multiple neurons in complex patterns. We each have about 89 billion neurons, and every one can connect with up to a thousand others. The potential for forming memories is enormous. One study, from Stanford University, calculated that the average human brain has the equivalent of 2.5 million gigabytes of computer memory.

The comparison with computers is useful but limited, because human memory is both selective and changeable. Most of the information that goes into the brain is ignored or rapidly junked. The bits that are retained are those that had the greatest impact emotionally.

Furthermore, two people never experience a shared event in an identical way, so each of us stores our own version. In computer terms, it is like opening a document to find that some of it has gone missing since it was written and other bits bear no resemblance to what went in originally.

Even consolidated memories – the ones that are transferred, over time, from the hippocampus to the cortex – are subject to alteration. Every time they are recalled, the firing patterns that recreate them can be overwritten or added to, so when they are reconsolidated – put out of consciousness – a slightly different version is stored.

Some memories are stored in unconscious parts of the brain. This makes them less susceptible to change. Procedural memories (learned body movements), for example, are stored in areas that control automatic movement. When neurons in this area are activated, they don't readily form new connections with the cortical neurons producing current conscious thought. So, if you are riding a bike while thinking about the swim you just had, you would, thankfully, have to visualise the swimming action very strongly indeed before your legs would begin to flap rather than rotate.

MEMORY DISTORTION

In 1974 researchers Elizabeth Loftus and John Palmer, from University California, Irvine, showed participants a film of two cars colliding, then asked them to describe what they saw. Some were asked, 'How fast were the cars going when they crashed?' Others were asked the cars' speed when they 'collided'. Those asked about the 'crash' speed estimated that the cars were travelling significantly faster than the speeds given by those asked about a 'collision' speed. The study demonstates how memory can be distorted just by the words used to ask about it.

Are recovered memories real?

→ They can be. It's quite common to forget something – even something highly traumatic – until something else happens that triggers it back into consciousness.

If someone asks you if you can swim, you will be able to tell them whether you can, and if you know the capital of Peru, you will retrieve the fact during a pub quiz, or at least get that 'on the tip of the tongue' feeing that informs you that the information is in there somewhere.

Many memories, however, are not available to our conscious mind. Usually they relate to things that have little resonance with current experience. Your memories of a childhood holiday, for example, may be so unlike the things you experience now that the memory rarely gets reignited. It might only come back if you find yourself in exactly the same place, or see someone you met there. 'Of course,' you say, 'I remember now!'

'Recovered', however, usually refers to the sudden resurrection during talking therapy of a memory of something nasty that happened in childhood. The accuracy of such memories is contentious, because a memory can be grossly distorted but feel entirely real. Nevertheless, some recovered memories accurately reflect past events, and may affect a person's behaviour even when they can't be recalled.

If a person is very scared by something, their amygdala, the tiny limbic system nucleus that responds to perceived threats, becomes sensitised to anything that suggests the scary event is going to happen again. If such an experience happens to a child before they can speak, their memory does not, obviously, have words attached to it and, as conscious thoughts are mainly couched in words, thinking will not bring the experience to mind; something else within the memory – an image or a smell – might, however. Someone who was terrified by a dog as an infant might remain fearful of dogs all their life, for instance, even though they may not know why.

POST-TRAUMATIC STRESS DISORDER (PTSD)

Memories may be buried in the brain and only recollected when triggered. Traumatic events can be stored in the amygdala – the unconscious brain module that creates fear. PTSD is often accompanied by such memories, which might consist of the feeling of terror and sensory associations (the smell of burning, for example, or a particular type of touch). Therapy can help the person recall the entire memory, in which case it can be consciously 'reframed', helping to reduce its emotional impact.

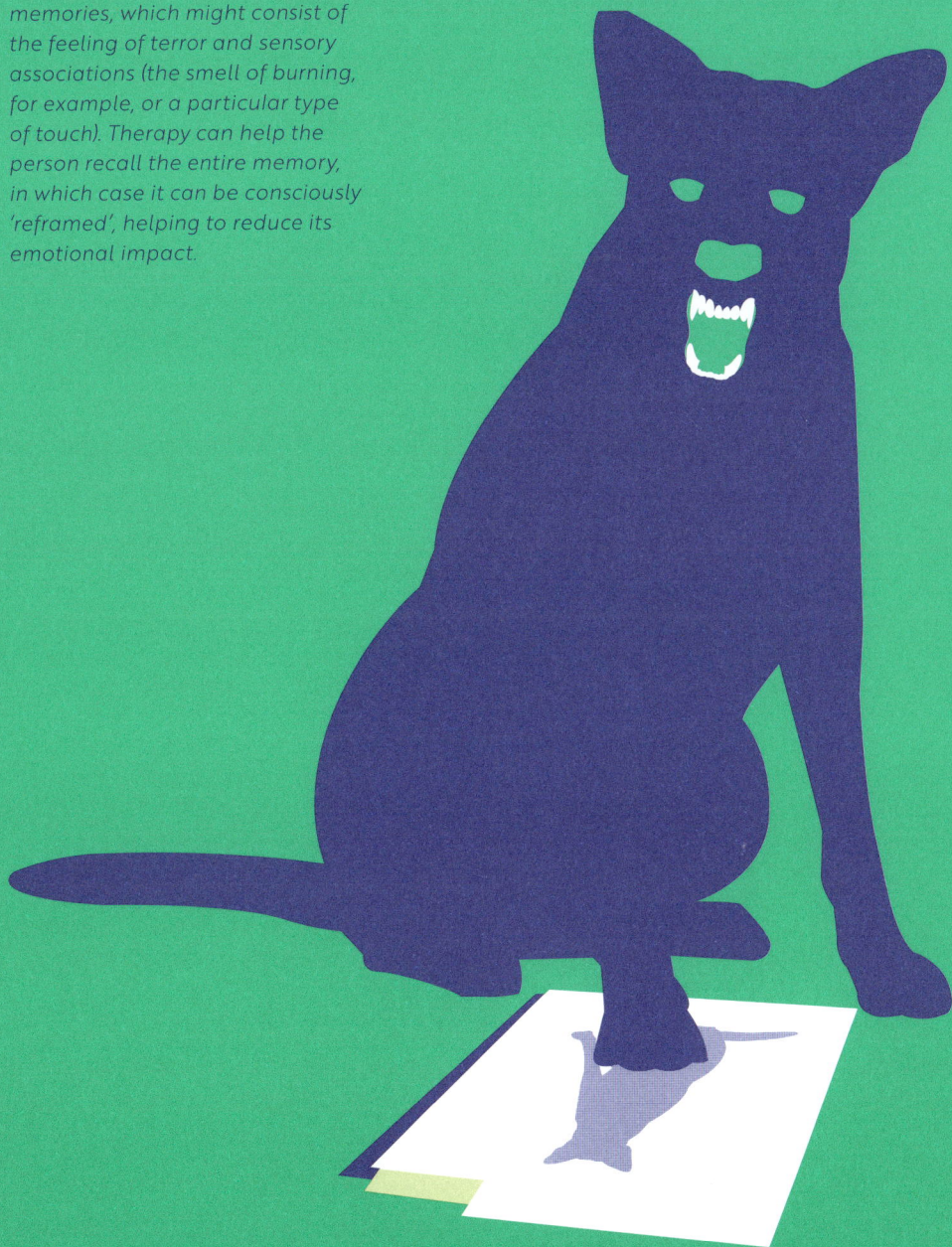

Why does memory deteriorate in old age?

→ Most people develop what's known as age-related memory dysfunction in later life, even though they are healthy. This may progress to dementia, in which brain tissue dies, but not necessarily.

There are many reasons that older people tend to have worse memories than the young. One is deteriorating health: the brain, like all body organs, is liable to get less efficient at repair and maintenance. Blood vessels tend to narrow and stiffen in old age, so oxygen and nutrient delivery to the brain – the hungriest organ we have – is reduced. There is also an increased risk of stroke, including mini-strokes, which may hardly be noticed at the time but slowly nibble away at brain tissue. Older brains also tend to accumulate debris – protein clumps which cut off oxygen supply to neurons and cause them to die. Alzheimer's dementia is associated with large quantities of this debris, but most people have a certain amount by the time they are 60 years old.

Memory formation also depends strongly on sensory impact. The strongest memories are those that involve multiple modalities, such as smell, taste, sound and sight as well as thoughts and emotions. Senses tend to dim with age simply because the sense organs deteriorate.

Another reason is that older people tend to be less socially, intellectually and physically active. Excitement and novelty help brains to retain experiences as memories, and older people may be less emotionally excitable and therefore much less likely to encounter new experiences.

The good news for older people is that, even if recalling specific facts or events may become slower and less accurate, the accumulation of memory-based knowledge tends to improve their ability to make 'wise' decisions – they have more life experiences to guide them.

The way to maintain good memory in to old age is to keep generally healthy, seek out new and challenging things to do, and to work at laying down memories through deliberately attending and recalling those things worthy of note.

KEEPING MEMORY ALIVE

Deliberately triggering recall by looking at old photos, visiting old haunts and seeing people from the past can refresh childhood memories in old age. A long-term study, reported in The British Medical Journal in 2023, of nearly 30,000 older people determined six lifestyle factors associated with maintaining a good memory: eating healthily, regular exercise (a daily 30-minute walk), social activity (at least two interactions per week), intellectual challenge (card games, crosswords, at least twice a week), no smoking and no alcohol.

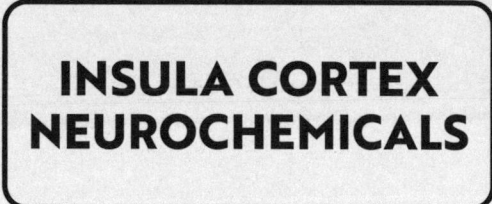

CHAPTER 5

EMOTIONS

FEELINGS

ANTERIOR CINGULATE CORTEX

ANIMAL BEHAVIOUR

THEORY OF MIND

INTRODUCTION

I f Spock-like aliens examined a hoard of cultural artefacts from planet Earth, they would be baffled. Our books, poetry, music and visual art are nearly all designed to represent or evoke emotion, of which Vulcans know nothing. If the aliens observed the creatures who produced the artefacts they would probably be even more mystified, just as Spock often was by the behaviour of the human *Star Trek* crew.

Perhaps if Vulcans had evolved on the constantly changing, perilous planet Earth, they too would know what it is to love and feel scared, angry, disgusted, happy and sad. Emotions, like everything our brain does, evolved to help us survive in an environment that speedily killed off creatures who were unable to adapt to it.

Many of the challenges that faced our distant ancestors (and still face most animals) called for physical action, such as flight or fight, and the prospects of surviving were enhanced in those who could do it quickly. It helped, too, to have some mechanism that automatically made their bodies reject or avoid harmful substances such as poisons. When they suffered setbacks and losses on account of straying into dangerous ground, it was probably useful to be robbed of energy, so they laid low, and when they were in a good place, it would have been beneficial to become active and exploratory. These physical adaptations are, basically, **EMOTIONS**.

They are orchestrated by our **AUTONOMIC NERVOUS SYSTEM (ANS)**, an unconscious control system which regulates heart and respiration rate and many other bodily functions to ensure that our physical state matches the demands of our environment. One branch of it, the sympathetic ANS, prepares us for action, and another, the parasympathetic ANS, returns us to a normal resting state. This balancing act is called **HOMEOSTASIS**.

No doubt all of this would make perfect sense to Spock and his friends, but they would still not understand why emotions feature so heavily in our books and songs. They would no doubt conclude there must be something more to it, and, of course, there is: **FEELINGS**.

This chapter looks at how the brain turns a physiological survival system into the feelings we recognise as emotions. It describes why women seem more emotional than men, and asks whether animals feel things as we do. Does everyone feel the same emotions? And what actually counts as an emotion: Sex? Pain? Are **MOODS** or **URGES** emotions?

EMOTIONS MAP

CONTROL

AUTONOMIC NERVOUS SYSTEM (ANS)
Unconscious control system that regulates bodily functions to ensure that our physical state matches the demands of our environment.

PARASYMPATHETIC NERVOUS SYSTEM (PSNS)
Unconscious control system that restores normal bodily state (homeostasis) after a change of state produced in response to a real or imagined environmental challenge.

EMOTIONS
Physiological changes designed to prepare the body for a challenge or opportunity; may be consciously experienced as a feeling.

FEELINGS
Conscious awareness of emotions.

INSULAR CORTEX
Largely hidden part of the brain, tucked between the temporal and frontal lobes, that monitors the state of the body; sensitive to bodily peril and involved in pain and disgust.

AFFECTIVE DISORDERS
Mental and behavioural disorders linked to mood disturbance, including depression and anxiety.

ANTERIOR CINGULATE CORTEX
Front part of the deep groove that divides the hemispheres; crucial to emotional and attention regulation; sensitive to one's own errors, emotional memory, pain and self-monitoring.

CORTEX

DOPAMINE
Neurotransmitter that produces pleasurable anticipation.

OXYTOCIN
Hormone and neurochemical associated with bonding and caregiving.

URGES
Strong, urgent motivational pressure to achieve a specific goal. Once it is achieved, the brain usually releases serotonin, which produces a feeling of satisfaction.

SEROTONIN
Neurotransmitter linked to calmness and satisfaction.

MOODS
Long-lasting mindsets that are not easily changed by events (as compared to emotions) and which affect behaviour and worldview.

EMOTIONAL REGULATION
Complex process that keeps emotional activity controlled and sensitive to context, e.g., damps down amygdala activation to prevent panic.

THEORY OF MIND
Intuitive understanding and recognition of one's own and other people's minds and mental states; develops incrementally through infancy and is usually first noticeable from the age of three.

NEUROTRANSMITTERS

Why do I get butterflies?

⟶ **Emotions are physiological changes designed to prepare your body for a challenge or opportunity. Butterflies, along with shaky hands, sinking stomachs and weak knees, happen when we become conscious of those changes.**

Although we tend to use the words interchangeably, 'emotions' and 'feelings' are two separate things. Emotions are the continuous fluctuation of things like blood pressure, heartbeat, breathing and muscle tension, which keep our bodies ready to deal with whatever crops up. Most of the time we are unaware of it, but if something happens that suggests, for example, danger (a loud noise nearby) or anticipates pleasure (the sight of someone we love), the impact on our body may produce feelings.

Some people are very sensitive to emotions and feel them quickly and strongly. Others are less sensitive. People may even be entirely unaware of having an emotion.

Feelings are initially sensational – butterflies and so on. But once the brain is conscious of an emotional feeling, it seeks the cause and attends to it. If it is the sight of a dog that produced the feeling, for instance, the brain will scour its memory for what a dog is likely to do, based on prior experience. The thought of a dog bite might occur, and after that the thought of running away. The person may even find they are running already, because strong emotion can produce automatic actions before a person is even aware of the emotional feeling.

The body's repertoire of responses is fairly limited, and identifying one emotion from another is not always easy. The Capilano Suspension Bridge study (from 1974, when gender sensitivities were less considered) sent conspicuously attractive young women to interview male passers-by on two bridges. One was a wobbly suspension bridge, the other more stable. The interviewers offered their phone numbers at the end 'in case the males had any questions'. Of the men who later called the women, four fifths had walked across the more precarious suspension bridge. The researchers, Don Dutton and Arthur Aron from the University of British Columbia in Vancouver, suggested that either the men mistook the excitement of their risky endeavour for sexual attraction, or fear itself generated sexual interest.

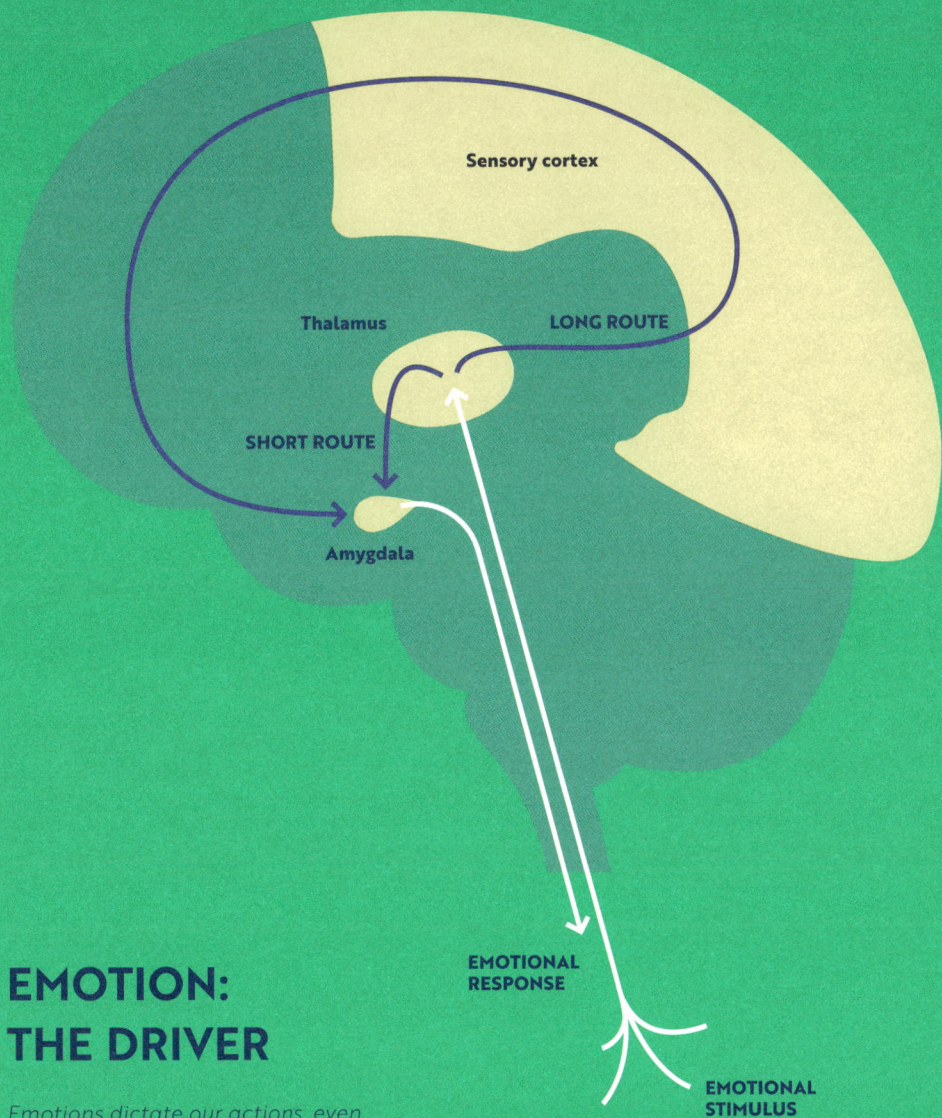

Sensory cortex

Thalamus

LONG ROUTE

SHORT ROUTE

Amygdala

EMOTIONAL
RESPONSE

EMOTIONAL
STIMULUS

EMOTION:
THE DRIVER

Emotions dictate our actions, even if we are not conscious of them. If we see something that makes us scared, for instance, the information is sent to the thalamus, then on to the cortex, which thinks about it before creating a conscious feeling via the amygdala and instructing a considered response. Meanwhile, the information also takes a shortcut from the thalamus direct to the amygdala, which generates a reflex reaction. This may be carried out before the cortex has registered a conscious feeling.

Does everyone feel the same things?

⟶ **Some emotions seem to be universal – fear, anger, disgust, joy, surprise – but people express them to a greater or lesser degree, and may feel them in response to different events.**

Emotions are pretty much the same in everyone: danger or challenge quickens the heart, increases blood pressure and breathing, and shunts blood to muscles in preparation for flight or fight. Disgust makes us move away and narrow our eyes and nostrils to avoid ingesting something nasty. Sadness produces tears – a call for support from others, and surprise forces open your eyes and mouth to take in information. Joy produces a flush of bodily warmth and an urge to move forward.

Similarly, in the brain each emotion is a distinct pattern of activity. The amygdala is active in anger, fear and, in the left hemisphere only, joy. Disgust activates the insula cortex, a largely hidden part of the brain tucked between the temporal and frontal lobes. These activities release neurochemicals such as dopamine (anticipation), noradrenaline (excitement) and serotonin (satisfaction). In many cases, two or more emotions are present at the same time.

Strong emotions generally produce conscious feelings, but it's difficult to know if these are the same for everyone, because all we have to go on is what people report. Someone with a sad-seeming body and a sad-looking brain (were we able to see inside it) might nevertheless say they feel amused or angry. Physiological disgust, in particular, is often described as something else. This may be because we often overwrite disgust with learned behaviour – valuing the perceived sophistication of strong cheese, for example.

The confusion might be cultural: there are societies in which negative feelings are socially unacceptable and in some groups it may be uncool to admit to fear or disgust. People who are habituated to danger are unlikely to feel it in a dangerous situation as strongly as someone who has rarely been at risk. Conversely, a person who constantly attends to their emotional state will probably feel danger more keenly and identify it more accurately.

MAPPED EMOTION

Every emotion has a physiological signature which is similar for everyone. Finnish researchers at Aalto University asked 700 volunteers to describe exactly which part of their body was affected by various emotions and came up with a body 'map'. Similarly, the brain produces a distinct activation pattern in each emotional state. A person's interpretation of their emotional state may differ, however, and people can find it difficult to identify what they are feeling, partly because most emotions are mixed. Behaviour may be more telling than words.

Anger   Fear   Disgust   Happiness   Sadness   Surprise   Neutral

Anxiety   Love   Depression   Contempt   Pride   Shame   Envy

Is sexual desire an emotion?

→ It's a grey area. Sexual desire on its own is more like an appetite or an urge – the body's attempt to hijack the brain to do something that benefits it.

Emotions and urges are similar in that they both stem from body states that are designed to generate action of some kind. The distinction between them is that urges are clearly directed at achieving a specific goal, *now*! When the goal is achieved, the brain removes the urge by ending the stimulus, and it may produce neurochemicals such as endorphins or serotonin, which activate the brain's pleasure areas.

Sexual desire, at least in most people, is much more complicated than the simple urge to eat or defecate. Scientists using brain imaging such as functional magnetic resonance imaging (fMRI) and electroencephalography (EEG) have pinpointed several brain regions that become active when people feel sexual desire, and some of them are concerned with sophisticated thought processes.

It takes about half a second for a person to recognise whether they are sexually attracted to someone and during that time many parts of the brain become active. The amygdala generates emotion, and the hippocampus, which encodes and retrieves personal memories, becomes busy –

possibly because it is searching past sexual experiences to help predict what action to take or expect. The anterior insula, the front of the deep infold between temporal and frontal lobes, monitors the body for changes. And frontal regions that make a person aware of what someone else is thinking and feeling also come into play.

All of this activity ramps up if sexual desire leads to sexual activity. Dopamine – the neurotransmitter that produces pleasurable anticipation – keeps the pleasure circuits humming, and oxytocin produces a 'bonding' emotion along with uterine contractions in women. During orgasm, parts of the prefrontal cortex that control behaviour are dialled down, while the anterior cingulate cortex – the front part of the deep groove that divides the hemispheres – turns on, blanking out pain to leave only pleasure.

THE URGENT BRAIN

Many urges arise because the body needs maintenance, such as feeding or voiding. Signals from cells lacking sugar, perhaps, or stretched by a full bladder enter the brain via the spinal cord. They go first to the thalamus, which speeds them to the relevant part of the cortex, like a traffic controller waving a wailing ambulance through a busy junction. The cortex turns the signals into familiar sensations such as hunger or the need to urinate, then it's over to the motor areas to create the actions necessary.

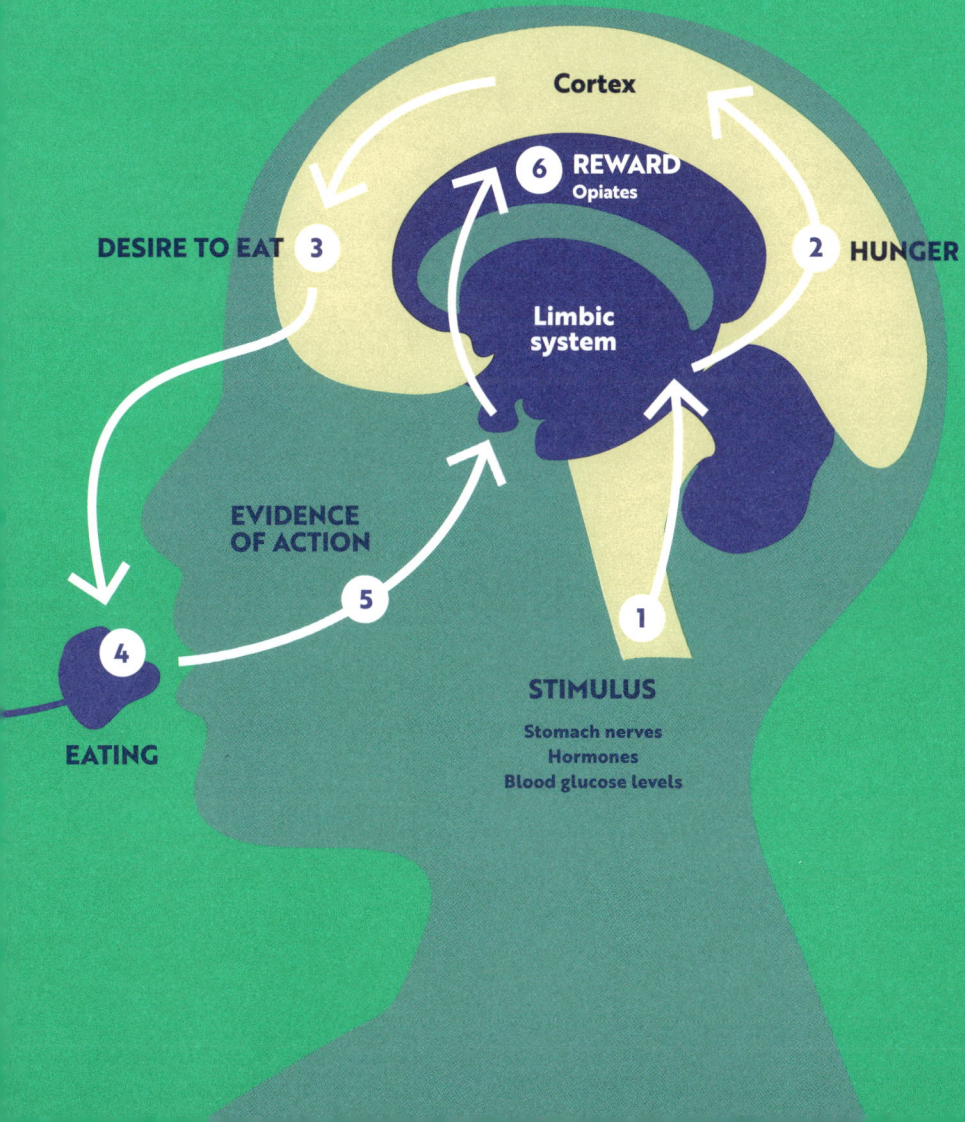

Cortex

6 REWARD
Opiates

DESIRE TO EAT 3

Limbic system

2 HUNGER

EVIDENCE OF ACTION

5

4

1

EATING

STIMULUS

Stomach nerves
Hormones
Blood glucose levels

Are moods just long-lasting emotions?

→ **There's more to them than that. Moods are mindsets – a sort of mental wallpaper – which are much less easily changed by events.**

The major distinction between emotions and moods is that moods last much longer than emotions. They tend to last for hours or days (longer than that and they may be diagnosed as affective disorders) and be less intense. Sometimes a mood might be blamed on a specific experience but more usually it is difficult for a person to pinpoint a cause. Emotions, by contrast, last seconds or minutes, dominate a person's subjective experience and are usually linked to a specific stimulus.

Compared to emotions, there is little neuroscientific research on the brain activity associated with the normal range of moods. Most of what we know is derived from studies of people who are depressed (characterised by chronic negative mood), anxious (fearful without due cause) or manic (in super-high spirits). In these cases there are clear abnormalities in brain activity and sometimes in anatomy, though it is difficult to know if physical changes are the cause or effect of prolonged mood-states.

People who are depressed or anxious, for example, show increased amygdala reaction in response to things like a sad face or a filthy room. Those with depression show less hippocampal activity when asked to remember something nice that happened to them; indeed, they may find it impossible to retrieve any personal memory of pleasure, and information exchange between parts of the frontal lobes and the rest of the brain is blunted. This pattern is associated with inward rumination rather than attention to the outside world. In manic people the pattern is reversed.

Although moods do not propel people to do particular things, they affect behaviour and may even alter your view of the world – literally. Observing a crowd of people, an anxious person's brain may pick out those who look aggressive; a depressed person may see frowns; a manic person may see smiles.

NATURE AS MOOD ENHANCER

Living in a natural environment rather than a city is known to have health benefits, among them the promotion of better mood. Studies show that mood is elevated by natural sights, smells and even the sensation of touching grass rather than concrete. People in natural environments have lower-frequency brainwaves in frontal areas – a pattern of activity that is a marker of comfort and relaxation. Urban environments, in contrast, appear to induce heightened amygdala activity, which is associated with anxiety.

Can animals feel things?

⟶ We don't know. But nor do we know for certain that other people can feel; our only guide is behaviour. Animals often behave as though they can feel, sometimes even more intensely than people.

When your dog wags its tail as you arrive home, its seems unarguably pleased to see you. But what about a fish, an octopus or a cockroach? The more an animal's behaviour resembles our own, the easier it is to assume that it feels like us too. Most mammals act as though they feel fear, anger, pleasure and pain, and studies of their brains show they have more or less the same emotional mechanisms as us.

Animal research has shown that many creatures have faculties long thought to be the sole preserve of human beings. Theory of mind, for instance – the (misleading) name for our automatic recognition that other people have minds of their own – has been demonstrated in dozens of species, including birds and possibly fish. Ravens hide food more quickly if they know they are being watched, and goats eschew food that is visible to a more aggressive goat, presumably because they know the dominant animal

would be displeased, and possibly aggressive, if it saw them snaffling it. Even fish seem to know when others can help them find food.

Many animals also show high levels of empathy. Chimps, for example, rush to comfort family members who have been hurt, throwing their arms around the victim and cuddling them just as people would. They also show anger, recognition of unfairness, and simple kindness. In a study carried out by the Dutch animal behaviourist Frans de Waal (b. 1948), two chimp friends were placed in adjacent cages while an experimenter repeatedly gave one of them grapes, which chimps love, and rather boring cucumber chunks to the other. The one receiving cucumber soon got fed up and chucked the cucumber back at the experimenter. The other chimp, seeming to recognise the unfairness of the situation, started to pass some of its own grape treats through the bars to its chum.

RAT EMPATHY EXPERIMENT

In 2011 researchers at the University of Chicago put pairs of rats in cages, with one free to roam while the other was held prisoner in a clear tube. The tube had a door that could be nudged open only from the outside. Most free-roaming rats soon learned how to release the door to set their captive companion free, and did so when there was no apparent advantage to the rescuer, suggesting that rats feel sorry for less fortunate mates. Rats' empathy probably evolved because it gives them a survival advantage in social situations, as it does in humans.

Are women more emotional than men?

→ Men and women's brains are physically different, on average, in a way that suggests women have more access to emotion. The difference may be biological, the result of learned behaviour or, probably, both.

Tradition dictates that women are more emotional than men, and countless psychological studies seem to back it up. In every culture where emotional behaviour has been rigorously measured, women report more intense feelings, more crying and more vivid emotional memories than men.

There are differences between male and female brains that suggest women may be better equipped than men to know their emotions. Their brains have more connections between the right hemisphere (where 'negative' emotions are generated) and the left hemisphere (where they are made conscious and describable).

Being able to describe an emotion is not the same as feeling it, however. For an emotion to be felt, the amygdala (in the unconscious limbic system) needs to send strong signals to the frontal lobe, where it becomes conscious. Both sexes respond to emotional stimuli by amygdala activation, but human brains can damp down the effect of the activity through a process called emotional regulation (ER). This happens, for example, if a person needs to be clear-headed in a dangerous situation, or when an angry outburst would be inappropriate.

Recent brain imaging research shows that men and women do this in different ways. Women's brains inhibit emotional response by sending signals from the frontal cortex to the limbic system, metaphorically throwing cold water on heated emotion. Men, by contrast, divert attention from the emotion, using circuits located in the back of the brain. Effectively, they prevent emotion from heating up in the first place. The male damping strategy is automatic, whereas women have to work quite hard, consciously, to keep emotion under control.

So, in some ways women are more emotional than men. But, given that emotions are primarily bodily changes rather than feelings, it remains possible that men have as much emotion as women – they just don't feel it.

TYPICAL FEMALE BRAIN

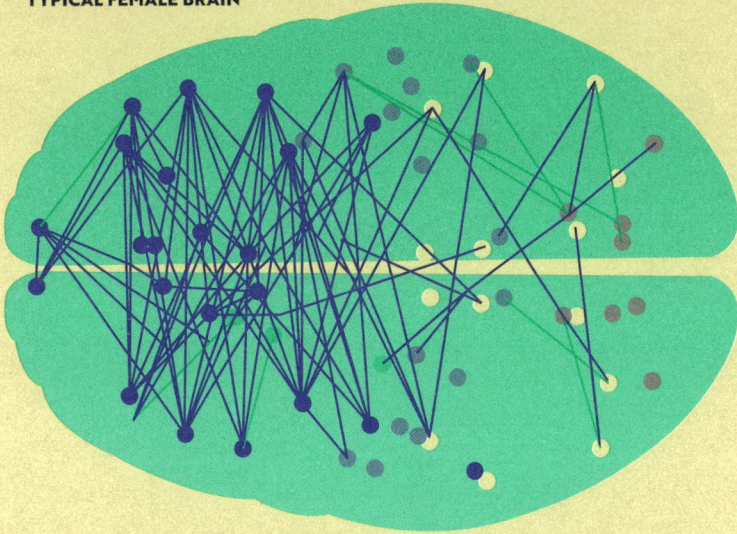

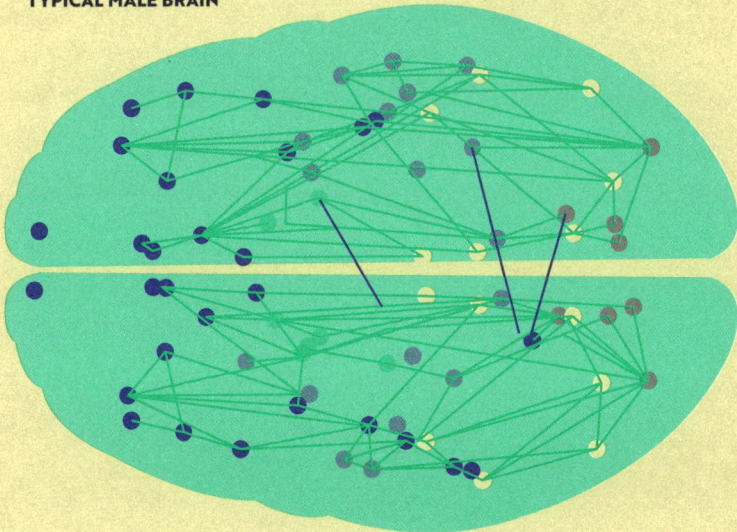

TYPICAL MALE BRAIN

CONNECTOME MAPS

Women's brains are wired differently from men. Brain imaging studies show that on average they have more neural pathways connecting the two hemispheres. Negative emotions are generated in the right hemisphere, and must be transmitted to the left before they can be expressed in words. Women may, therefore, have greater ability than men to feel emotions consciously and describe them.

SEX AND GENDER

TRANSCRANIAL DIRECT CURRENT STIMULATION

CHROMOSOMES

NEURODIVERSITY

SEXUALISATION

MATURITY PRINCIPLE

CHAPTER 6

DIVERSITY

IDENTICAL TWINS

PERSONALITY TYPES

THE BIG FIVE TRAITS

TESTOSTERONE

ADOLESCENCE

INTRODUCTION

Every brain is different from every other, and for this reason no two people experience the world or behave in exactly the same way. This is true from the moment we are born, because human brains are exquisitely sensitive to the environment. Even **IDENTICAL TWINS** are different from each other as babies because their brains are affected by the particular way they lie and move in the uterus. By the time we are adults we will each have experienced billions of unique events, and each one will have left some impression, however minute. Nevertheless, most of us remain similar enough to have common interests, more or less understand one another and move though our shared social world with moderate ease.

The exception to this is a significant minority of people known collectively as neurodiverse. Until about 2000, anyone who behaved in a distinctly different way from most others was considered to be eccentric (usually not in a good way) or medically disabled. This changed after Australian sociologist Judy Singer coined the term **NEURODIVERSITY** to describe people who, like herself, saw the world differently but were not in any way 'ill'.

The term 'neurotypical' was invented shortly after. It made its first appearance in 2010 on a spoof website for the fictional Institute for the Study of the Neurobiologically Typical, where it was defined as a 'disorder characterisation by preoccupation with social concerns, delusions of superiority and obsession with conformity'. Today its satirical origins have been lost and the word is used without irony to describe everyone who does not identify as neurodiverse.

Dividing the world into these two categories is socially useful because it has reduced the stigma that people frequently suffered if they were notably different and has drawn attention to the strengths, as well as weaknesses, that go with it. It can be confusing, though, because the line between neurotypical and neurodiverse is fuzzy and shifting.

Neurodiversity embraces people with identifiable physiological abnormalities such as Down's syndrome and cerebral palsy, through conditions like **DYSLEXIA** and dyscalculia, to **ATTENTION DEFICIT HYPERACTIVITY DISORDER (ADHD)** and **TOURETTE'S SYNDROME**. Neurodiversity is not a medical disorder (though some of the conditions it covers are) and according to some, one in three of us could be said to be 'on the spectrum'.

This chapter considers the entire range of human diversity and how it is measured. From attempts to measure **PERSONALITY TRAITS** to charting the seismic changes in the brain throughout **ADOLESCENCE**, it looks at some of the most contentious social issues of the day, including whether a person can have the visible physiology of one **SEX**, and the brain of another.

DIVERSITY MAP

SEX
The physical sex of a person. All foetuses start female and become male, if they have a Y chromosome, in response to baths of testosterone produced by the mother. If the testosterone fails, they retain female physiology.

SEXUALISATION
Staged process by which a person acquires sexual characteristics through the action of sex hormones. Physical sex, sexual identification and sexual orientation are determined at different stages and may not be consistent.

NEURODIVERSITY
Collective term for conditions in which the brain consistently produces notably untypical behaviour. There are three types: applied, clinical and acquired.

TESTOSTERONE
Male steroid sex hormone that moulds the body into the male form; also sexualises the brain during development, though not necessarily in the same way as the body.

CHROMOSOME
Long thread of DNA that carries genetic information, including a pair of sex chromosomes (XX – female; XY– male).

MATURITY PRINCIPLE
Idea that people, on average, become more extraverted, emotionally stable, agreeable and conscientious as they grow older.

IDENTITY

GENDER
A person's complex interrelationship between body, identity and social gender. Gender identity can match or differ from the determined sex at birth.

PHRENOLOGY
Popular theory and practice in the early 19th century centred on the notion that bumps on the skull were caused by brain modules that created particular aspects of personality.

GENES

IDENTICAL TWINS
Non-scientific term for monozygotic twins. Although genetically identical, their brains diverge rapidly due to moulding by environmental factors, even before birth.

AUTISM
Clinical neurodiverse condition that results in impaired social interaction and communication, and restricted and ritualised patterns of behaviour.

DYSLEXIA
Impaired ability to read and/or spell. Some dyslexic conditions are associated with visual pathway abnormalities.

TOURETTE'S SYNDROME
Clinical neurodiverse condition by which people make repeated involuntary movements or utterances (tics).

ATTENTION-DEFICIT HYPERACTIVITY DISORDER (ADHD)
Clinical neurodiverse condition characterised by persistent difficulties in concentrating, hyperactivity and impulsivity.

TRANSCRANIAL DIRECT CURRENT STIMULATION (tDCS)
Non-invasive neuromodulation that stimulates the brain directly with electrical current or magnetic pulses to change the flow of information between neurons.

TRAITS

BIG FIVE
(Or the Five Factor Model.) Detailed analysis of words used to describe human personality, with five fundamental dimensions: Openness, Conscientiousness, Extraversion, Agreeableness and Neuroticism.

Do identical twins have identical brains?

⟶ Although they share the same genes, identical twins have different brains, and therefore different personalities. Even at birth their brains will be distinct due to minutely different experiences in the womb.

Psychological studies of twins, especially pairs who have been brought up separately, have shown that genes have an astonishingly wide effect on a person's behaviour.

The Minnesota Twin Family Study, which began in 1989, has followed 81 pairs of identical twins who were separated at birth. The study has found that as adults, many of them share personality traits and behaviours that you would hardly guess were produced by protein-making molecules, our genes. A pair who were raised apart from the age of four weeks, for example, discovered when they reunited at the age of 39 that they both bit their nails, smoked the same brand of cigarette, drove the same type of car and vacationed at the same beach resort. Other twins have met after decades apart to find they are dressed almost identically, have the same haircut, do similar jobs and share a particular sense of humour. More rigorous observation has shown they have very similar IQs and personality profiles.

Yet identical twins are never identical in behaviour, even when raised together. Their mothers can identify behavioural distinctions between them within days, and some monozygotic twins (from the same egg) grow up to be almost as different in personality as any other siblings. The differences seem to be due to the brain's exquisite sensitivity to its environment. Work by Cambridge neuroscientist Simon Baron-Cohen has shown that even the position of a twin in the uterus can alter the brain. Girls who share the womb with a male twin tend to be more tomboyish than most girls, even when they have been brought up without their brother. The reason is thought to be that they share the pre-birth testosterone 'baths' that male foetuses trigger, and which sexualise boys' brains before birth.

Even newborn twins' brains look different. A recent study found that trained observers could detect anatomical differences between identical siblings very soon after birth, although in most cases they could pair up the twins just from looking at their brains.

SAME GENES, DIFFERENT BRAINS

Twin studies suggest that IQ and other measurable brain functions are almost exactly due equally to genes and nurture. Very tiny environmental factors can, however, lead to huge differences between identical siblings. One foetus may occupy a slightly advantageous position in the womb, for instance, and be stronger at birth. The developing brain is so sensitive that the slightest differences in prenatal nutrition (due to their position in the uterus, for example) can produce striking differences in later behaviour.

Is there a set number of personality types?

→ Some personality tests categorise people into types, but these are very rough-and-ready guides to what someone is like. A more precise way to gauge personality is to assess the 'Big Five' traits.

The Big Five personality traits are Openness, Conscientiousness, Extraversion, Agreeableness and Neuroticism. A lengthy Q&A test combined with complex mathematical calculations is used to determine how high or low an individual scores in each. Unlike tests that simply clump people together in broad types, this Five Factor Model (FFM) allows for an almost infinite range of personality.

Personality tests have been used for decades for vocation testing and psychological assessments, but recently the FFM has been used by neuroscientists to determine the brain anatomy and function that is responsible for personality distinctions.

Colin DeYoung, of Minnesota University, measured various brain regions using magnetic resonance imaging (MRI) in volunteers whose personalities had been assessed using the FFM. He found that people who scored highly on the five personality measures had visibly greater volume in the parts of their brains associated with those behaviours. In other words, it seems

that different personalities have visibly different brains. In theory, we might one day be able to assess a person's personality just by looking at that bit of their brain.

Other work has shown that personality is also reflected in the chemicals produced by the brain. Extraverts make more dopamine, for example, the chemical known to produce the feeling of pleasurable anticipation, and people who score high on neuroticism (marked by a tendency to anxiety and depression) have been found to have chronically high levels of the stress hormone cortisol. Openness (which includes imagination, curiosity and inventiveness) is also associated with high dopamine, but in the brain circuits associated with attention rather than just those linked to reward.

Brain research on personality is still in its infancy, but it seems likely that one day the neural basis of behaviour may be understood well enough for personality to be artificially moulded, just as plastic surgery can change our bodies.

THE NEW PHRENOLOGY

Phrenology is the idea – hugely popular in the early 19th century – that bumps on a person's skull were caused by brain modules that created particular aspects of personality. Phrenologists performed 'skull readings' to determine their clients' characteristics.

Ironically, the demise of phrenology was largely due to the discovery of real brain modules – and neuroscientists have found that the size of certain brain areas can predict personality traits. Rather than a porcelain head, today's researchers use sophisticated medical imagery.

Does your personality change over time?

⟶ The brain undergoes massive neural reorganisation through infancy and adolescence, and personality often changes dramatically as a result. From adulthood onwards, personality is usually fairly stable, give or take the odd bit of life turmoil.

Children's brains undergo enormous changes in the first three years as neuronal pathways are formed, pruned and reformed until, like a sculpture emerging from stone, their personalities become clear. After that, a child's personality is fairly settled until adolescence, when most young people undergo sudden, often dramatic and seemingly antisocial personality changes.

Teenagers have to contend with massive destabilising pressures brought on by social and sexual development, and on top of that their brains undergo a fundamental reorganisation. The change is caused, in both sexes, by testosterone, which makes neural pathways exceptionally plastic, so connections make and break easily. This allows them to learn new things and forget old ones, very quickly, and is reflected in sudden changes in friendships, hobbies, tastes and behaviours. Already formed ideas are likely to be rejected and new ones tried out, then just as suddenly dumped in favour of something else. Emotions run high.

The frontal lobes, meanwhile, are not fully 'online', because the neural pathways in them are still not completely sheathed by myelin – the fatty substance that allows electricity to pass efficiently through brain tissue. In mature adults, fully myelinated frontal lobes prevent excessive behaviour by inhibiting impulses, but without this top-down control, personality traits such as risk-taking and pessimism are amplified. In teenagers this may lead to problems such as dangerous or criminal behaviour. The good news is that once the frontal lobes are fully functioning – usually around the age of 25 – an individual's personality is more or less settled. Thereafter, change is usually subtle and gradual, unless the person undergoes a dramatic life event or brain injury.

Even better news is that long-term studies of personality suggest that there is a tendency for it to improve over time. People become more extraverted, emotionally stable, agreeable and conscientious as they grow older. Psychologists call it the 'maturity principle'.

TEEN ANGST

Teenagers are vulnerable to mental disorders because their brains are subject to massive reorganisation around adolescence that creates temporary instability. Bourgeoning sex hormones soften normally stable neural connections, rocking the mental scaffolding that supports an individual's beliefs, habits and identity. During this time, a person is easily distracted and misled, and a tendency towards pessimism or worry may speedily develop into depression or anxiety. Additional external challenges such as the Covid pandemic and bullying on social media increase the risk of destabilisation.

Do neurodiverse people have different brains from others?

→ Everyone has a brain that is different from everyone else's. 'Neurodiverse' is just an umbrella term for people whose brains produce notably untypical behaviour.

Even the most ordinary individual occasionally does or says something that is not entirely typical of normal people. To that extent, all of us are neurodivergent. Some people (about one in five by current estimates), however, consistently behave so unusually that they are labelled with a neurodivergent condition. There are hundreds of such labels – even homelessness gets an entry in the latest Diagnostic and Statistical Manual of Mental Disorders.

Neurodivergent conditions are generally classified as applied, clinical or acquired. Applied neurodiversity is marked by difficulty in one or more cognitive areas, such as language, motor control or arithmetic. Examples are dyslexia, dyspraxia and dyscalculia. They are not considered to be health conditions. Clinical neurodiversity includes attention deficit hyperactivity disorder (ADHD), autism spectrum condition (ASD), Tourette's syndrome and intellectual disability. These are considered to be disabilities. Acquired neurodiversity embraces neurological conditions that affect behaviour or experience and which are associated with illness, injury or psychological trauma. They may be temporary or permanent.

Brain imaging studies show that almost every type of neurodivergence is associated with abnormal brain anatomy or function. 'Abnormal', however, does not necessarily mean 'bad'. It just means that a person's brain is wired up in a way that does not equip them to function optimally in normal society. People with dyslexia, for example, find reading difficult, so they may perform poorly in our word-based education system, whereas they may do brilliantly in a system that focuses on visual or auditory teaching. People with autism usually struggle to function in a social milieu that requires intuitive responses to emotion, yet those same people may have great talent in other areas. Even those who have had brain injury due to stroke can benefit from their affliction: one man completely lost his characteristic pessimism after a right-hemisphere stroke and was left in irrepressibly good humour.

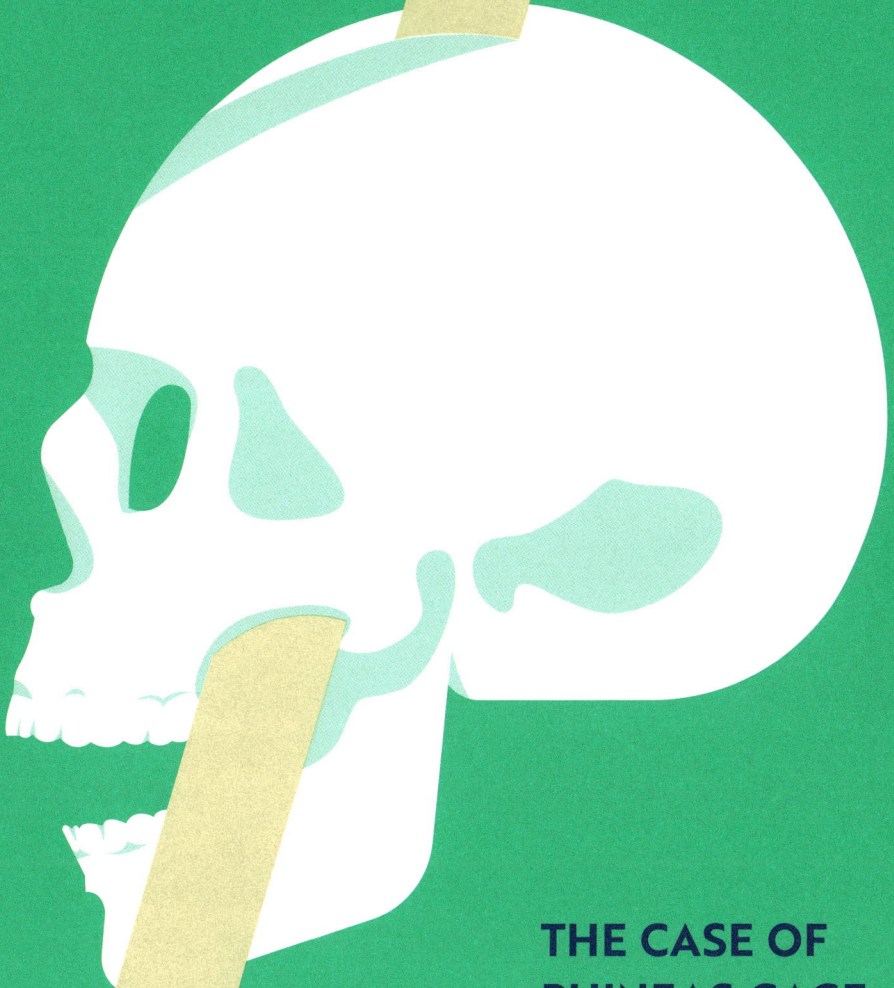

THE CASE OF PHINEAS GAGE

In 1848 a 25-year-old US railroad worker called Phineas Gage (1823–1860) had his brain refashioned dramatically when a steel rod went through the front of it during a mistimed explosion. After the event, Gage, formerly a conscientious, polite and conventional man, was reported to behave with abnormal abandon, diverging dramatically from social norms. His case is still widely cited today to demonstrate how fundamentally behaviour is dictated by the physical state of a person's brain.

Can neurodiverse people 'learn' typical behaviour?

→ Children with attention disorders often become typically attentive as their brains mature, and many neurodiverse people learn to mimic neurotypical behaviour.

Neurodivergent behaviour may cause problems for the person and/or other people, even when it is not a health issue. People with dyslexia, for example, may be deprived of learning opportunities because so much knowledge is conveyed via the written word, and people on the autistic spectrum may long for closer social engagement but be excluded because they have unusual interests or behaviour. Children with attention disorders (particularly those with associated hyperactivity) may be punished for what may seem to be deliberately antisocial behaviour. In these cases it is beneficial for neurodiverse people to learn to behave more neurotypically, at least some of the time.

The extent to which this is possible depends on the person's type and degree of neurodiversity, and the help available to them. Applied neurodivergent conditions, where a person has difficulty with particular cognitive tasks such as reading or maths, can be minimised with specialised teaching.

The earlier this happens, the more successful it is. Those who have an identified brain abnormality may learn to 'act normal' if they can understand how they differ from others. People on the autistic spectrum, for example, may work out what other people are feeling by studying facial expressions and learning how to react.

Even Tourette's syndrome, in which people make repeated involuntary movements or utterances, can sometimes be controlled. Tics occur when a person's frontal lobes fail to inhibit powerful motor signals. Frontal lobe function is ramped up, however, during focused attention, so if the person learns to concentrate, the tics may stop. Many musicians, artists and even surgeons with Tourette's syndrome behave normally while working.

Acquired neurodiversity, the sort caused by brain trauma or injury, may be improved by therapy. If left-hemisphere language areas are affected, for example, the equivalent areas in the right hemisphere may be taught to take over from them.

CHANGING THE BRAIN

Some types of neurodiversity can be altered by stimulating the brain directly with electrical current or magnetic pulses. Transcranial direct current stimulation (tDCS) passes a tiny electric current into the brain through the skull, where it changes the flow of information between neurons. Direct stimulation has been found to improve most applied neurodivergent conditions and many clinical and acquired types, including movement and language problems, attention disorders and some aspects of autism. This non-invasive technique is well researched but remains little known and rarely used.

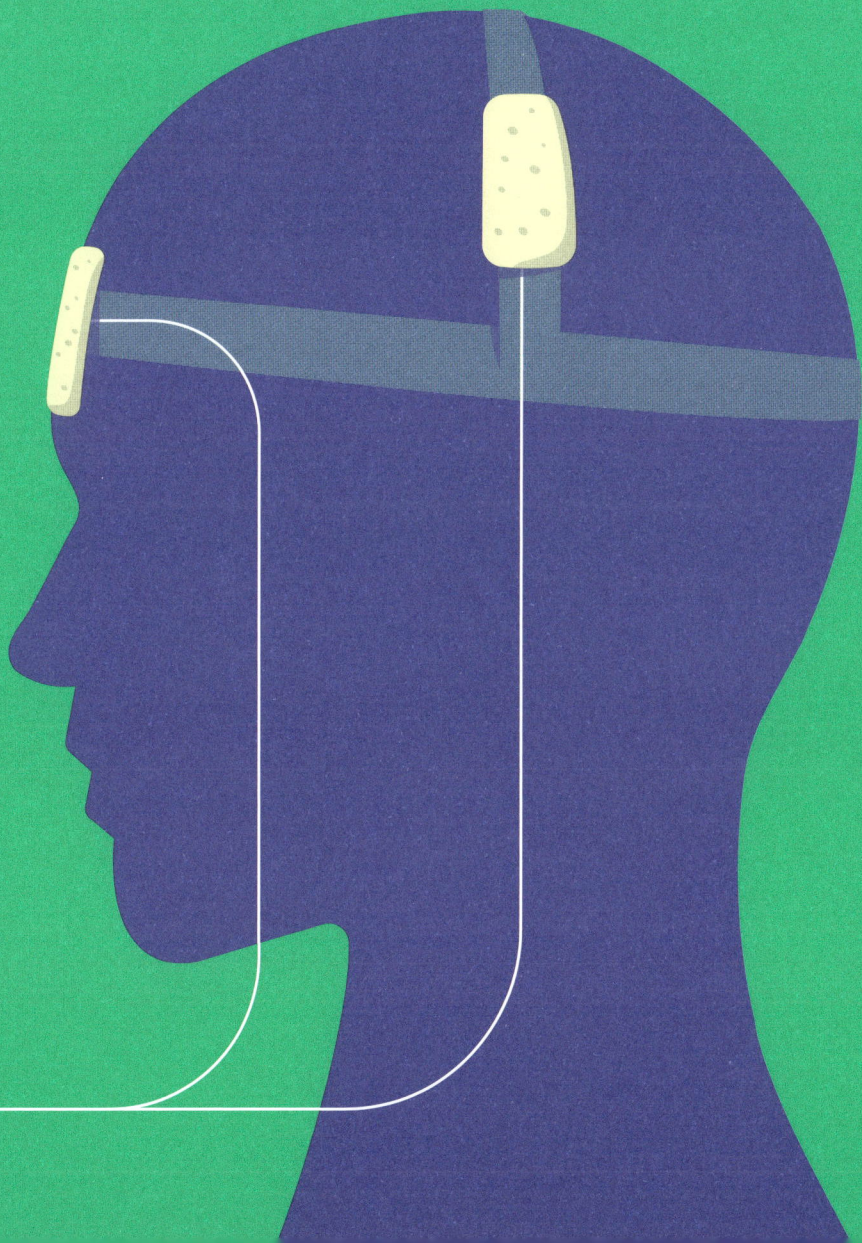

Can my brain be a different sex to my body?

➜ It is possible. Physical sex and gender identity are largely determined before birth by a complicated hormonal process triggered by genes. Sometimes the body is sexualised in one way and the brain another.

All babies start developing along lines that, without intervention, would make them female. However, if a foetus has the typical male XY chromosome rather than the female XX, its genes trigger the production of testosterone, which moulds the body into the male form. By week fourteen of gestation, the foetus has developed male gonads and genitals. Female foetuses do not trigger testosterone and instead develop ovaries at around eleven to twelve weeks.

Later in pregnancy, during the last trimester, boys' genes trigger further testosterone baths, which mainly affect brain tissue. Curiously, testosterone is turned into oestrogen in the foetus's brain, but its effects are to reorganise neural tissue in a way that will produce typically male behaviour in later life. Parts of the hypothalamus, amygdala and other (mainly limbic) structures are altered in such a way that they will be exceptionally responsive in future to hormonal stimulation. The amygdala, for example, is primed to generate aggressive behaviour in the face of challenge, rather than appeasement, and the hypothalamus is primed to enhance libido. More subtle changes produce the inner sense of being male, known as male gender identity. The effects of this pre-birth 'masculinisation' become particularly clear during puberty, when the child's body is again flooded with male hormones.

The fact that body and brain sexualisation are two separate processes means that one can occur without the other, creating a male body but a non-masculinised (and therefore a feminine) brain. It is also possible that a female foetus's brain can be masculinised *in utero*. Little is known about how this happens. Some studies suggest that the late-pregnancy testosterone bath may be suppressed in pregnant women who are particularly stressed or take certain drugs.

Some of the effects of pre-birth sexualisation can be reversed in later life. Generally, it is easier to change bodily, rather than mental, sexual characteristics.

THE SEX SPECTRUM

According to some research, such as that by Swaab and Garcia-Falgueras at the Netherlands Institute for Neuroscience, physical sex and gender identity do not fall into two distinct categories – male or female; rather, we all lie somewhere on various spectrums of sexual identity. Body sex is created by robust genetic processes, but about 1.7 per cent of babies are born with visible features of both females and males. Two or three times as many people are estimated to have minds – which probably means brains – that do not produce stereotypically male or female experience and behaviour. The Genderbread Person (developed by Sam Killermann) is a popular model for understanding the complexity of gender.

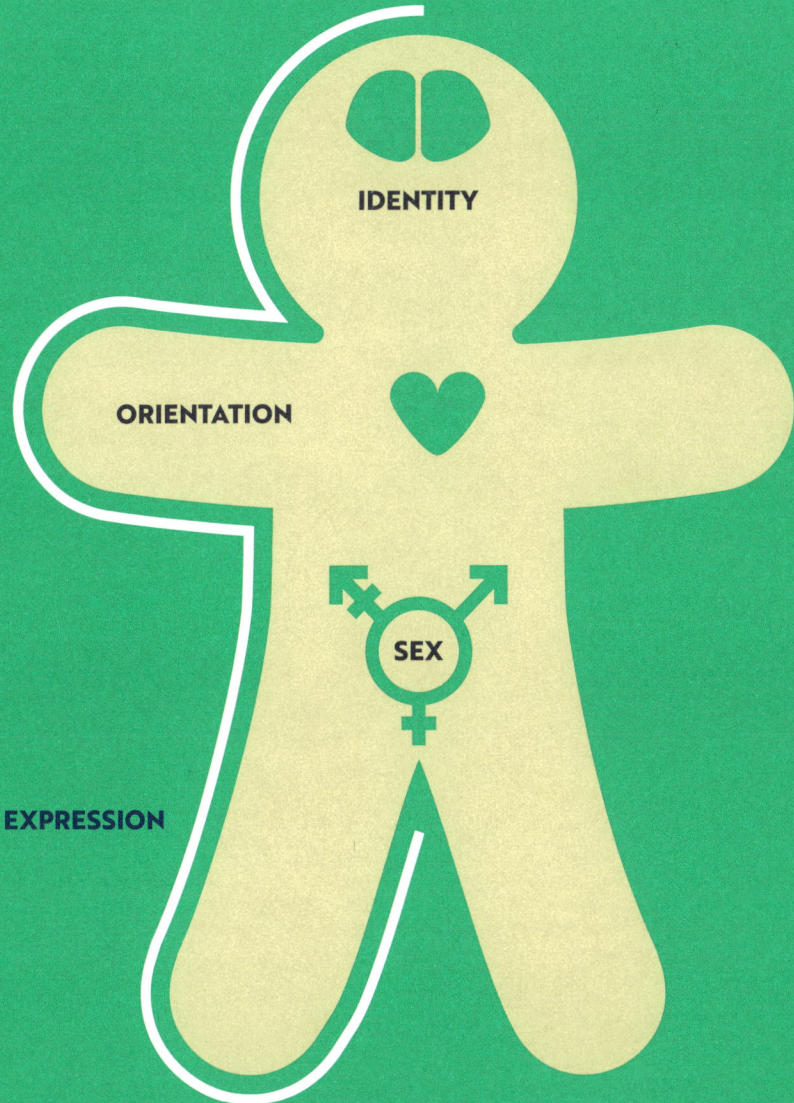

IDENTITY

ORIENTATION

SEX

EXPRESSION

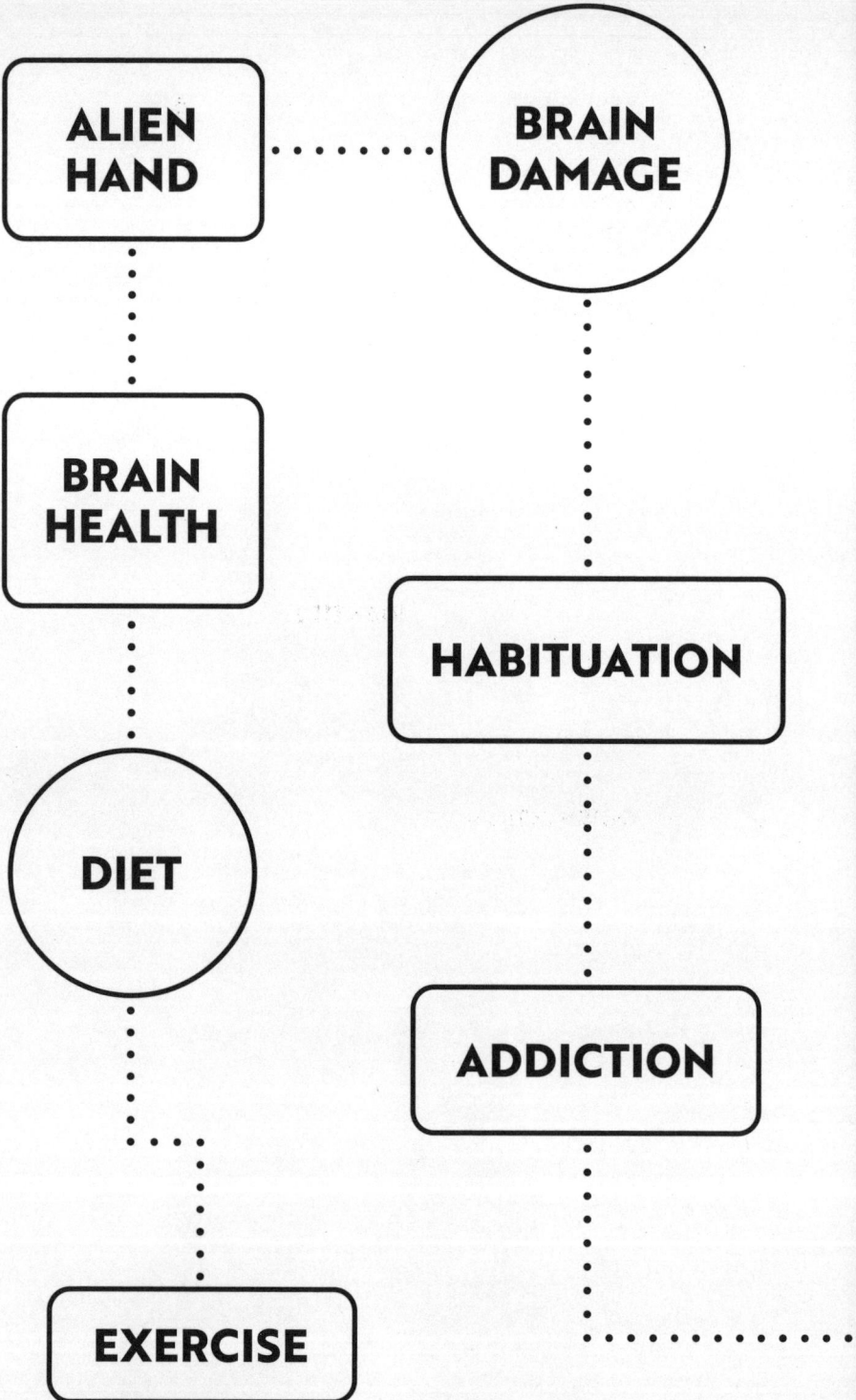

CHAPTER 7

BRAIN FITNESS

MENTAL HEALTH

DEMENTIA

ALZHEIMER'S

DRUGS

NOOTROPICS

INTRODUCTION

Conscious experience is so unlike anything else in the world that it's easy to forget that it relies on a physical organ. Our ability to perceive, understand and remember depends on the smooth functioning of the brain, and we need to keep it physically healthy just as much as any other part of our body.

There's plenty we can do to protect our brain and help it work well for our lifetime. Good food, exercise and the right sort of mental stimulation will go a long way in preventing **STROKE** and the form of **DEMENTIA** caused by mini-strokes. Even **ALZHEIMER'S DISEASE** can be staved off to some extent.

Fit brains do not necessarily generate continuous feelings of light and joy. Your brain is designed to adapt to the world and sometimes unpleasant things like worry, sadness and tiredness abound. All it means is that, in response to an emergency or abnormal situation, your brain is forcing you to think through things you would rather ignore by making you worry, or demotivating you so you rest up. When the situation is back to normal, your brain should get back to normal too.

Sometimes, though, a brain gets stuck in emergency mode long after it's useful. Long-term stress, a major loss or a big social setback, for example, can rob the brain of its ability to bounce back and lead to **MENTAL HEALTH** conditions such as **DEPRESSION** or **ANXIETY**. Such conditions should not be ignored, because, if untreated, they will eventually cause physical changes that will further reduce the brain's ability to return to normal. Most conditions will respond to talking therapy and/or drugs such as anti-depressants. Taking a pill to relieve mental discomfort might seem strange, until you remember that it is a physical malfunction producing the feelings, so it makes sense that a physical therapy may relieve them.

This chapter suggests how to assess whether brain glitches, like forgetfulness, may be indicative of something more serious such as stroke or dementia. It looks at whether substances known as **NOOTROPICS** or recreational drugs can make your brain work better, and also at **ADDICTION** and its impact on the brain. It also considers the effect of brain damage on personality – the extent that physical injury alters a person's behaviour and whether such changes can be reversed. Finally, it outlines the everyday habits, including **COGNITIVE EXERCISE**, that can help keep your brain healthy.

BRAIN FITNESS MAP

HABITUATION
Second component of addiction: having to consume something to maintain homeostasis.

MENTAL HEALTH
Emotional, psychological and social well-being; embracing cognition, behaviour and perception.

TOLERANCE
Having to have or do more and more of something in order to get the same stimulus or effect (as in addiction).

COGNITIVE EXERCISE
Mental tasks and activities to use and strengthen the brain through forging new neural connections and increasing cerebral blood flow.

NOOTROPICS
So-called 'smart drugs' that claim to improve intellect or cognitive performance.

DEPRESSION
Sustained state of sadness and pessimism, with symptoms such as lethargy and sleep and appetite disturbances. Untreated, it can make it increasingly difficult for the brain to generate pleasure.

MU RECEPTORS
Opioid receptors that cause euphoria and help regulate the body's response to pain.

HEALTH

ADDICTION
Repeatedly and/or continuously doing something you associate with pleasure even when it has a negative impact on your life; has three components: craving, habituation and tolerance.

ANXIETY
Heightened apprehension or fear; if not justified by the situation – i.e., the threat does not objectively exist – it may be considered a disorder.

DISEASE

STROKE
Brain cell death caused by blocked or broken blood vessel. Small strokes are common but in the long term cause dementia; major strokes are life-threatening and need emergency treatment.

DEMENTIA
Impairment of memory and cognition caused by brain tissue degeneration; becomes increasingly common with age.

ALZHEIMER'S DISEASE
The most common form of dementia, linked to a build-up of proteins in the brain, which affects how cells transmit messages.

AMYLOID PLAQUE
Clumps of protein pieces, which can block cell-to-cell signalling at synapses and activate immune system cells that trigger inflammation and devour disabled cells; a hallmark of Alzheimer's.

TAU TANGLES
Abnormal forms of tau proteins that accumulate in tangles inside neurons; a hallmark of Alzheimer's.

GAMMA-AMINOBUTYRIC ACID (GABA)
Neurotransmitter that inhibits neuronal excitability to maintain normal cognition.

NEURONS

Do mental health problems indicate brain disease?

⟶ Probably not. The gamut of mental health disorders are caused by unusual neural activity that can occur in perfectly healthy brains. That said, damage to the fabric of the brain can trigger some mood disorders.

Your brain evolved to keep you alive, not to make you feel good. Unpleasant mental states such as severe anxiety, sadness and fear are feelings brought about by emotions, essentially body changes, that once helped people survive in very dangerous situations.

Today there is less call for strong emotions – we rarely have to attack or run away from predators – but the brain keeps producing them. Some people's brains are particularly prone to this because of genetic factors or experience, or usually a combination of the two. These people are extra-vulnerable to constant rumination, lack of motivation, panic attacks, obsessive behaviours, fear of abandonment and all the other 300-odd mental health problems currently listed in the psychiatrists' Diagnostic and Statistical Manual (DSM).

Simply feeling bad is not itself a disorder, but when feelings arise for no discernible reason, or if they persist after the initial cause has gone, or are so bad that normal life becomes impossible, they become disorders. The DSM and various other certified tests provide criteria for diagnosis. For example, the DSM states that to be diagnosed as clinically depressed, a person must have suffered a minimum of five symptoms, of which low mood is one, for at least two weeks.

Treatment for mental disorders is hit and miss, but it is important to seek help if you think you have a mental problem. If left untreated, some disorders can make life intolerable, and make it increasingly difficult to return to normality. Long-term depression, for example, changes the brain's anatomy. Imaging studies have found shrinkage in multiple brain areas with long histories of depression, particularly in the connective tissue in the brain's reward circuit and the hippocampus. Some of the change may be due to excess stress hormones, which also occur in chronic anxiety. Eventually the brain cannot respond normally to good things, thereby perpetuating the original problem.

IMAGING MENTAL HEALTH

Mental health problems such as mood changes and odd behaviour are sometimes due to damage to the fabric of the brain. The cause is usually an identifiable event such as a stroke or head injury, and may be visible by MRI, which shows the state of soft body tissue. 'Functional' disorders, in which the brain is healthy but working unusually, are sometimes visible as well. Reduced activity in the brain's reward system, for instance, is often visible in scans of a depressed person's brain .

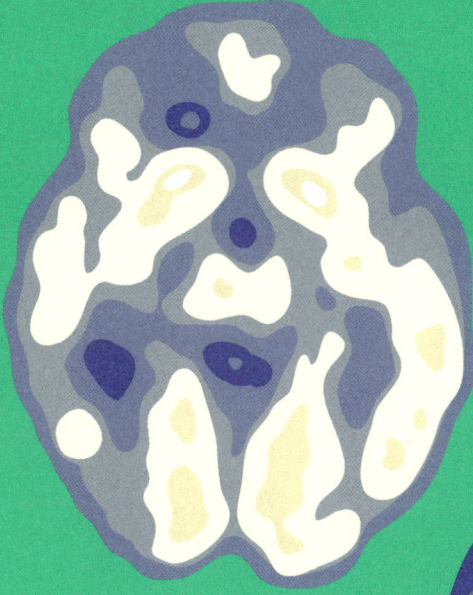

NOT DEPRESSED

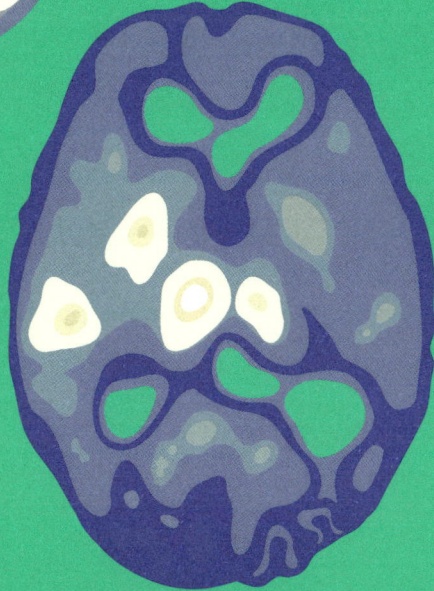

DEPRESSED

How can you tell forgetfulness from early dementia?

⟶ **With difficulty. The first symptoms of dementia can easily be ignored. People are usually diagnosed only when memory slips start to impact on everyday life.**

Occasionally forgetting where you put your keys is normal, and struggling to remember a name happens to everyone. Even walking into a room and wondering why you are there is quite common, especially as you get older. What, though, if things like this start happening more often? How do you know if it is the first sign of dementia?

There are several types of dementia, each of which has a slightly different pattern of symptoms. Memory is invariably affected, but the type of dementia that shows up first and most clearly as forgetfulness is Alzheimer's disease.

Memory problems that are persistent enough to cause problems beyond momentary irritation could signal its start. Other symptoms include: difficulties with making plans; an inability to complete familiar tasks such as making tea; confusion about times and places; and previously unknown problems with reading and writing.

It is possible to develop Alzheimer's in your thirties, but it is very unlikely. Forgetfulness at that age is more likely to be due to other factors such as stress, excessive fatigue, adverse drug effects or some other ailment altogether. By the age of 65, however, one in ten people show signs of Alzheimer's; and over the age of 85, one in three people are diagnosed with the disease. People carrying particular gene mutations are at greater risk than others.

Multi-infarct dementia, in which multiple tiny strokes pepper the brain with minute dead patches, may look much like Alzheimer's disease. It usually occurs mainly in the frontal lobes, and tends to cause personality changes at an early stage.

You might think that a disease with known genetic risk factors and an apparently clear cause and course would be easy to spot and treat. Unfortunately, it is not. Brain scans and blood tests can reveal protein accumulation and drugs may slow the disease, but at the moment there is no cure.

ALZHEIMER'S DISEASE

Alzheimer's is a progressive disorder of the brain. It is characterised by the build-up of amyloid and tau proteins that become toxic to the brain. Tau accumulates in tangles inside neurons, while amyloid forms plaques that collect between neurons. Gradually, neurons lose the ability to communicate and cells die. Parts of the brain shrink, beginning in the hippocampus, causing problems with memory.

HEALTHY BRAIN

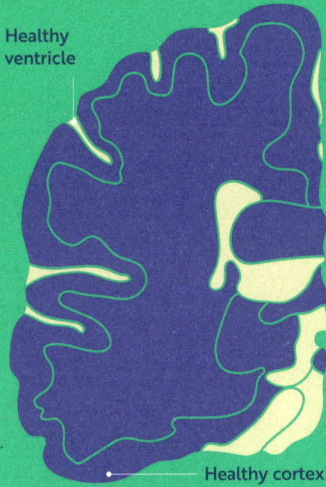

Healthy ventricle

Healthy cortex

HEALTHY NEURON

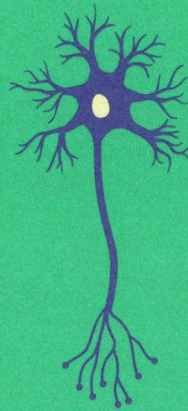

BRAIN WITH ALZHEIMER'S

Enlarged ventricle

Shrunken cortex

AFFECTED NEURON

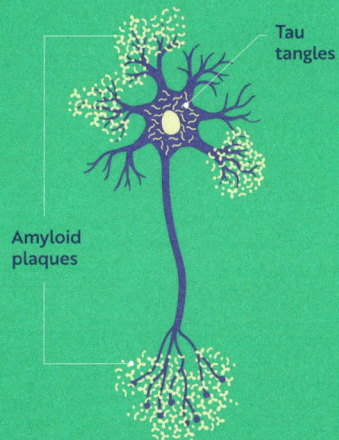

Tau tangles

Amyloid plaques

Can drugs make my brain work better?

⟶ Hundreds of substances claim to be 'smart drugs' or 'cognitive enhancers'. Collectively, they are called nootropics. If these substances did what they claim to, the world would be bursting with geniuses.

Nootropics include dietary supplements, medicines and recreational drugs such as cocaine and MDMA (ecstasy). Some are useless, some dangerous; others are legal when prescribed by a doctor but not when bought from a stranger on a dance floor. Many are cooked up and sold by dealers; new ones appear every week. The illegality of many drugs has curtailed research on them, so their risks and benefits remain largely unknown. Illegal drugs do not go through the rigorous safety check that pharmaceutical products must go through, and there is little or no quality control nor dose regulation, which makes recreational drug use pretty risky.

Most brain drugs work (or are claimed to work) by increasing or enhancing noradrenaline, dopamine, γ-aminobutyric acid (GABA) and/or serotonin. The effect of these neurotransmitters differs according to which part of the brain they act on, and drugs, similarly, produce differing effects depending on where they work. Cognitive enhancers mainly target the frontal lobes; hallucinogens target the sensory areas at the back of the brain; calming drugs work mainly on the limbic system.

Amphetamines, for example, increase noradrenaline, dopamine and serotonin in the frontal lobes, increasing alertness and focus. This type of drug is prescribed for children with attention deficiency hyperactivity disorder (ADHD), where they inhibit impulsive actions. In those without ADHD, they can boost cognition for a short time, but repeated use may create addiction, heart problems and psychiatric problems.

MDMA floods the brain's frontal lobes with dopamine, inducing a loved-up high for a few hours, but causing a corresponding low afterwards. Modafinil, a medicinal drug prescribed to people with narcolepsy and ADHD, also increases attention and alertness, and seems to have relatively few side-effects, but its use is illegal unless prescribed by a doctor. Some antidepressants boost cognition but others damp it down – it often depends on an individual's genetic make-up. Most of them take weeks to work and may need to be taken for long periods, or even for life, to avoid relapses.

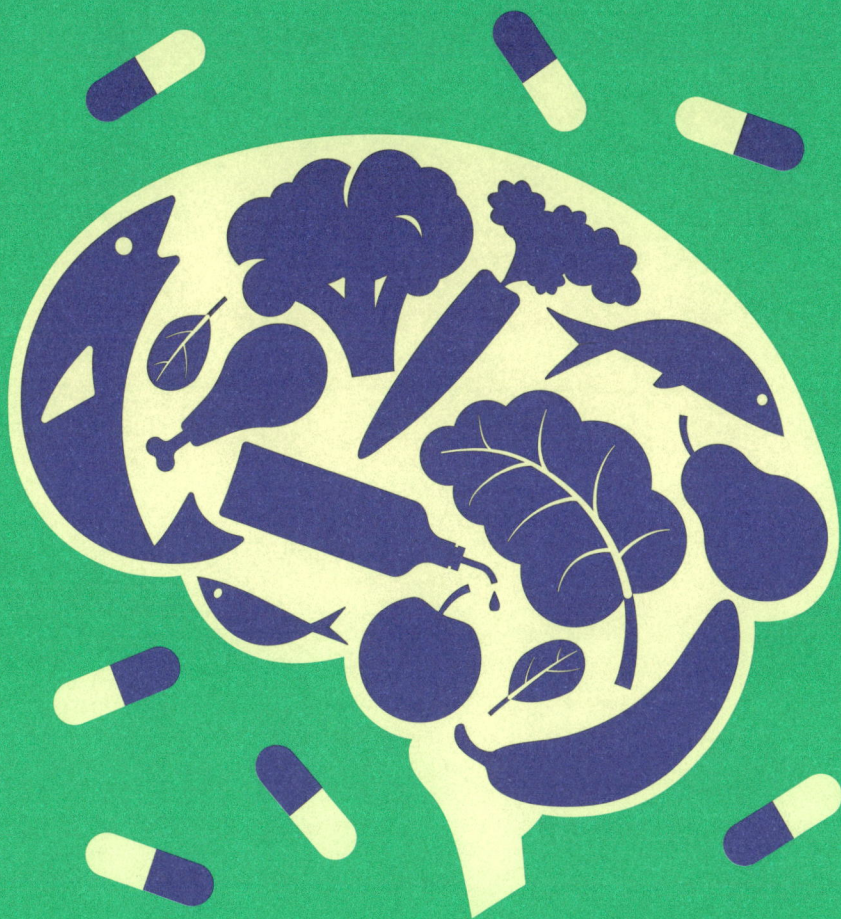

DIETARY SUPPLEMENTS

Dozens of dietary supplements are meant to enhance brain function but few of them have been rigorously tested in independent trials. Many of them are claimed to work by enhancing general health, especially heart and circulation. The brain certainly requires good nutrition and uninterrupted blood flow, but healthy people can get adequate vitamins, minerals, fish oils and similar brain nutrients by eating a diet rich in fresh fruit and vegetables, wholegrains, olive oil and white protein such as chicken and fish.

Can you become addicted to anything?

→ If it gives you pleasure, yes. You can get addicted to anything that – at first – creates a quick, big pleasurable chemical change in the brain. But some things prove to be more addictive than others.

Certain substances have a direct impact in the brain: cocaine releases a flood of dopamine, creating pleasurable excitement. Opioids such as heroin latch on to 'mu' receptors in brain cells, which cause euphoria; and alcohol, among other things, increases GABA, a neurotransmitter that inhibits anxiety. Other things achieve similar brain changes indirectly: shopping, sex, gaming. The crucial thing is the intensity and speed of reward. Gambling is most likely to become addictive when it involves quick-reward betting such as fruit machines. The brain links laying the bet and receiving the reward because they happen almost simultaneously, and the drive to repeat it is undimmed by disappointment. Many products and consumables – sugary snacks, for instance – are designed to give a near-immediate surge of pleasure.

Simple enjoyment ends and addiction begins when you continue doing something you associate with pleasure even when it has a negative impact on your life. Addiction can cause severe social problems; addiction to substances also damage your body. In short, addiction can interfere with your mental capacity for normal life and subject you to multiple health and social risks.

One effect of repeating something nice is that your brain dials down its pleasurable response. Your brain evolved to keep you alive, after all, and if you are continuously high you are probably not going to make the most life-preserving decisions. Constant hits of quick pleasure reduce the number of receptors that respond to pleasure-making chemicals. The result is what's known as habituation: in order to get that kick, you have to consume or do more and more of whatever it is that turns you on.

The reduction in receptors means that all pleasure is reduced, so normal life is flat and the 'down' that inevitably follows the 'high' may be severe. Your body may need the thing you are addicted to just to feel normal. Cravings may become so intense that soon nothing else matters.

SOCIAL MEDIA ADDICTION

Social media is increasingly recognised as potentially addictive. Professor Ofir Turel of California State University claims that up to one in five people in the US are addicted to internet technology. He found that the brains of people who use social media excessively show some of the same anatomical changes that are seen in those addicted to drugs, including reduced grey matter in areas that process reward. The finding suggests that their brains have downgraded their pleasure response.

Can brain damage cause changes to personality?

→ If part of your brain is put out of action, the behaviour associated with that part, and possibly every part, will be affected. The brain is very good at rewiring itself, though, so the alteration might be temporary.

Your personality is formed by your genetic inheritance together with a lifetime of experience. However, even a deep and consistent trait, such as patience or extraversion, can be altered in seconds by a severe stroke or knock on the head.

Most people who have strokes report feeling changed by the event. Apathy, anxiety, depression and irritability are most commonly reported, along with lack of interest in sex or (more rarely) heightened sexual appetite. The changes may be temporary and in some cases the brain rewires itself, reverting to previous ways of behaving.

The effect of brain damage depends on where it occurs and how bad it is. Every bit of the brain works in conjunction with every other bit, so the effect of an injury is rarely clear-cut. Generally, however, injuries that occur in the frontal lobes are most likely to affect judgement, decision-making, inhibition and emotion – aspects of cognition most closely associated with personality.

For example, if the stroke or injury is on the side of the frontal lobes (dorsolateral cortex), the person may lose interest in things they would normally enjoy and find it difficult to follow instructions or to understand subtle or abstract reasoning. Damage to the front (orbitofrontal cortex) may reduce inhibition, making a previously cautious and polite person impulsive and perhaps even offensive. They may lose control of their emotions, suddenly flaring up with bad temper or showing uncharacteristic lack of concern for other people. Lower down, in the ventromedial area, brain damage can produce unusual moral judgements. Asked to make the right decision when faced with the (theoretical) prospect of having to harm either strangers or loved ones, people with damage to this area have been found more likely than others to leave aside emotional factors and simply choose the option that benefits the greater number.

ALIEN HAND

One of the strangest behavioural problems that can occur after brain injury is 'alien hand', sometimes called 'Dr Strangelove syndrome', after the satirical film character played by Peter Sellers. One hand does an intentional action while the other immediately does the opposite. In one case, a man clasped his wife to kiss her with one hand then hit her with the other. In another case, a woman found that when she selected a piece of clothing from her wardrobe, her alien hand threw it to the floor.

How can I keep my brain healthy?

⟶ The first thing to do is to keep it physically well, just like any other part of the body. In addition, your brain will benefit from a special type of exercise.

The brain is a greedy organ that relies on a constant supply of blood to deliver the nutrients it needs. The most important thing you can do to keep it healthy is to ensure that nothing impedes the blood flow. That means looking after your heart and circulation, which in turn means eating a healthy diet and doing physical exercise.

Three diets have been shown to slow cognitive decline. The Mediterranean diet, based on the traditional cuisine of southern European countries, is rich in fruit and vegetables, whole grains, beans, nuts, seeds and olive oil. Fish, poultry and dairy foods are eaten sparingly, and red meat rarely. A rigid version of this, the DASH diet (Dietary Approaches to Stop Hypertension), has been developed to reduce blood pressure, and a combination of the two – the Mediterranean-DASH-Intervention for Neurodegenerative Delay, or, put more simply, the MIND diet – is designed specifically to protect the brain. It emphasises berries and green leafy vegetables such as spinach and kale.

The other important component of a brain-healthy lifestyle is exercise. Again, the sort that keeps your heart healthy is also good for the brain, and you don't have to run a marathon – just twenty minutes' brisk walking four days a week will keep the circulation flowing.

In addition to this, though, your brain needs cognitive exercise. That means engaging in social activities, learning new things and taking on mental challenges. The more it is used, the stronger the brain gets. When it is active, and particularly when it is learning new things, it forges new neural connections, bulking itself up like a muscle. Active neurons attract blood flow, so mental exercise helps to keep them alive, and if dementia does develop, the more brain tissue that has been built up, the more the disease has to destroy before the symptoms become severe.

BRAIN-HEALTHY LIFESTYLE

The single-most important way to protect your brain is to protect your heart. The Mediterranean diet keeps heart and blood vessels healthy, but there is no one magic ingredient. A moderate amount of red wine is fine, but alcohol is generally not good. Social engagement seems to be particularly beneficial for keeping minds lively, along with learning new things. Although new neurons are rarely made in adulthood, the connective tissue that carries information around the brain is increased by mental activity.

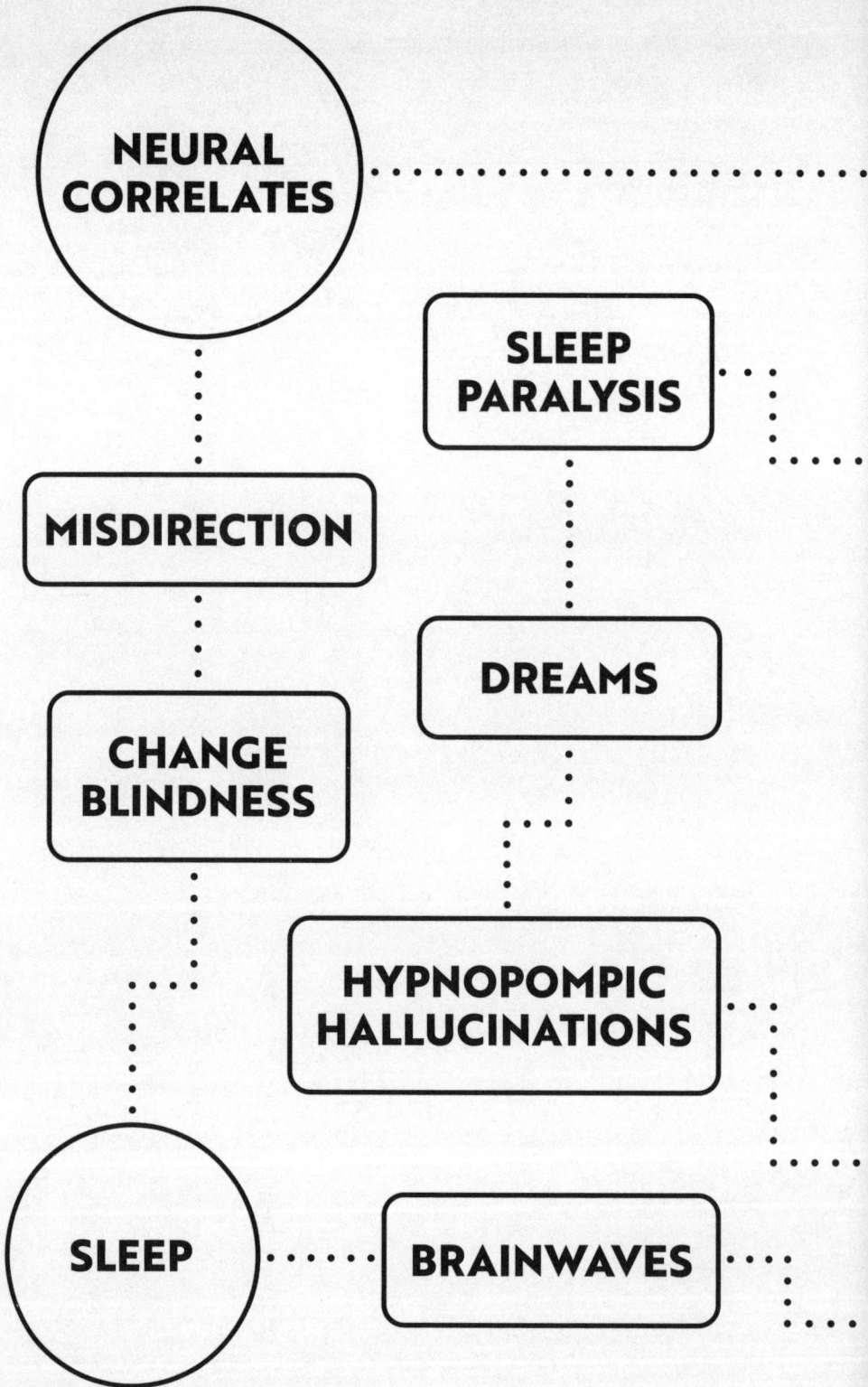

NEURAL CORRELATES

SLEEP PARALYSIS

MISDIRECTION

DREAMS

CHANGE BLINDNESS

HYPNOPOMPIC HALLUCINATIONS

SLEEP

BRAINWAVES

CHAPTER 8

CONSCIOUSNESS

- CONSCIOUS STATES
- ALTERED STATES
- MEDITATION
- ARTIFICIAL INTELLIGENCE
- HYPNAGOGIC ILLUSIONS

INTRODUCTION

I t's easy to ignore consciousness because we are completely immersed in it, like the air we breathe. If you turn your mind to it, though, you'll find it throws up intriguing questions.

Start with a seemingly simple one: What is it, this thing we call consciousness? The question, sometimes posed as the mind–body problem, has baffled people since records began and possibly even earlier. When philosophy and science diverged in the 19th century, both disciplines claimed the problem for their own, and a genteel competition to solve it continues to this day.

A quarter of a century ago, leading neuroscientist Christof Koch (b. 1956) made a bet with the (then) young philosopher David Chalmers (b. 1966). He wagered that by 2023 brain studies would have arrived at a clear and complete model of consciousness. Chalmers (whose own theory is that consciousness is an essential part of the material universe like time or space) said the mystery would never be cracked just by looking at the brain. Since then scientists have found many **NEURAL CORELATES** of consciousness – brain activity that marks **CONSCIOUS STATES** – but they are far from complete. Chalmers won his bet.

While scientists may not yet have cracked the 'hard' question of consciousness, they have delved deep into countless 'easy' problems and come up with some amazing revelations. Take another apparently straightforward question: Are you conscious, right now, of your surroundings? Now, keeping your eyes on these words, ask yourself some precise questions about the things around you. Is there anything on the floor? If so, exactly where and how is it lying? Is there a bookshelf nearby? Is it full? If there are curtains, can you describe the pattern, precisely? When you've answered these questions, look around. Chances are that you only thought you were conscious of the things around you. Even the simple matter of what we are aware of is much more complicated than it seems.

In addition, what happens when consciousness goes away, as in **SLEEP STATES**? How do we stay alive? And what about dreaming? Our eyes are closed, our body is paralysed (we might even be in **SLEEP PARALYSIS**), yet our mind takes us off on amazing adventures in which we see and feel things – through **HYPNAGOGIC ILLUSIONS** and **HYPNOPOMPIC HALLUCINATIONS** – as vividly as when we are awake. How can we be conscious of all these things when our sense organs are effectively switched off? And what if we seek an **ALTERED STATE**, through meditation or drug use – how does our brain respond?

This chapter considers these and other questions, and explores their strange answers.

CONSCIOUSNESS MAP

BRAINWAVES
Synchronised electrical pulses from neuronal assemblies that show how fast neurons are firing; measured using EEG.

NEURAL CORRELATES OF CONSCIOUSNESS
Brain activity patterns that occur when a person reports being conscious.

LUCID DREAMING
Waking up while dreaming but continuing to hallucinate the dreamscape. May include feeling of being paralysed, of someone being present or of being pinned down.

MICROTUBULES
Tiny protein tunnels that run through all cells, including neurons, and play a part in sculpting the shape of the cell and transporting essentials to sustain neuronal activity.

SLEEP PARALYSIS
A biological 'switch' that blocks motor signals from brain to body while we sleep.

METABOLITES
Waste products of metabolism. These can build up in the brain, contributing to the debris associated with certain types of dementia.

SLEEP STATES
There are four stages of sleep: N1, 2 and 3 and REM (rapid eye movement, or dream sleep). Each one is characterised by different types of brainwaves.

CEREBROSPINAL FLUID
Flows in and around the brain and spinal cord to protect from injury and provide nutrients; washes out metabolites while we sleep.

GLUTAMATE
Brain and nervous system's major excitatory neurotransmitter. In a normal brain it is kept within narrow limits: too much can cause nerve cells to die; too little prevents normal cognition.

ALTERED STATES

Abnormal form of consciousness, usually an ecstatic feeling outside the normal range of experience, which can be achieved through taking drugs or meditating, for example.

MEDITATION

Mental discipline in which the practitioner manipulates their attention – commonly on breath, a mantra, stillness or the present moment – to access an elevated state of consciousness.

TURING TEST

Test devised by mathematician Alan Turing to assess whether a computer can think, in which a person evaluates if they are interacting with a machine or a human.

DEFAULT MODE NETWORK (DMN)

Circuit that is activated in the brain when it is ruminating without a specific goal. The thoughts it generates has with one's self as the focus rather than the outside world.

HYPNAGOGIC ILLUSIONS

Brief, vivid images that pop up in consciousness as we fall asleep, sometimes replaying something we did earlier. They usually fade as we enter the N1 stage of sleep.

HYPNOPOMPIC HALLUCINATIONS

Often experienced when we emerge from a sleep cycle; similar to hypnagogic phenomena but tend to become more coherent, as the brain is speeding up rather than slowing down.

SLEEP

Why is consciousness the 'hard' problem?

→ Consciousness has a reputation for being difficult, because no one knows what it is or how it comes about. What is known, though, is some of what the brain does to produce it.

Consciousness is different from everything else we know. You can't touch it, see it, measure it or locate it, and until recently it was widely considered to be beyond scientific exploration. However, in the last 30 years scientists have been able to look inside conscious brains and see what is going on. Technologies such as fMRI, PET (positron emission tomography) and EEG (electroencephalography) allow researchers to observe neural activity and, by asking people to describe their experience while the observations are made, match the activity to many aspects of consciousness.

In 1995, when this sort of research was taking off, the Australian philosopher and cognitive scientist David Chalmers (b. 1966) described the search for the neural correlates of conscious experience as one of the 'easy' problems of consciousness. Other 'easy' problems include how the brain learns, produces emotions and makes decisions – all things that neuroscience has been, and continues to be, brilliant at revealing.

He pointed out that a 'hard' problem, however, remains: How does the material universe generate this apparently non-material thing?

The problem remains as tricky as ever. There are, though, many theories. Some people (including David Chalmers) maintain that consciousness is a fundamental part of the universe: 'the flip side of material'. Others think it is generated by super quantum computing in 'microtubules' – tiny parts of each neuron. Several theories hold that special features of brain anatomy or functioning are key, and others believe – as, famously, did the 17th-century philosopher René Descartes – that consciousness is some sort of 'spirit stuff'. In this view, the brain does not produce consciousness but acts as a conduit for it, much as a radio antenna picks up a distant broadcast. There are also people who think consciousness does not exist as something in its own right but that it is actually an illusion, and others who think it just 'emerges' from complexity, social interaction or computation.

CONSCIOUS STATES

Conscious experience, and its neural correlates, vary widely. When you are working towards a goal, for example, the pattern of information flow around the brain is almost exactly the opposite of how it is when you are freely daydreaming. Conscious states can be displayed on three axes: 1) Activity – how excited your brain is; 2) Selectivity – how narrow or wide your attention is; 3) Focus – whether you are generating your thoughts internally or looking to the outside world.

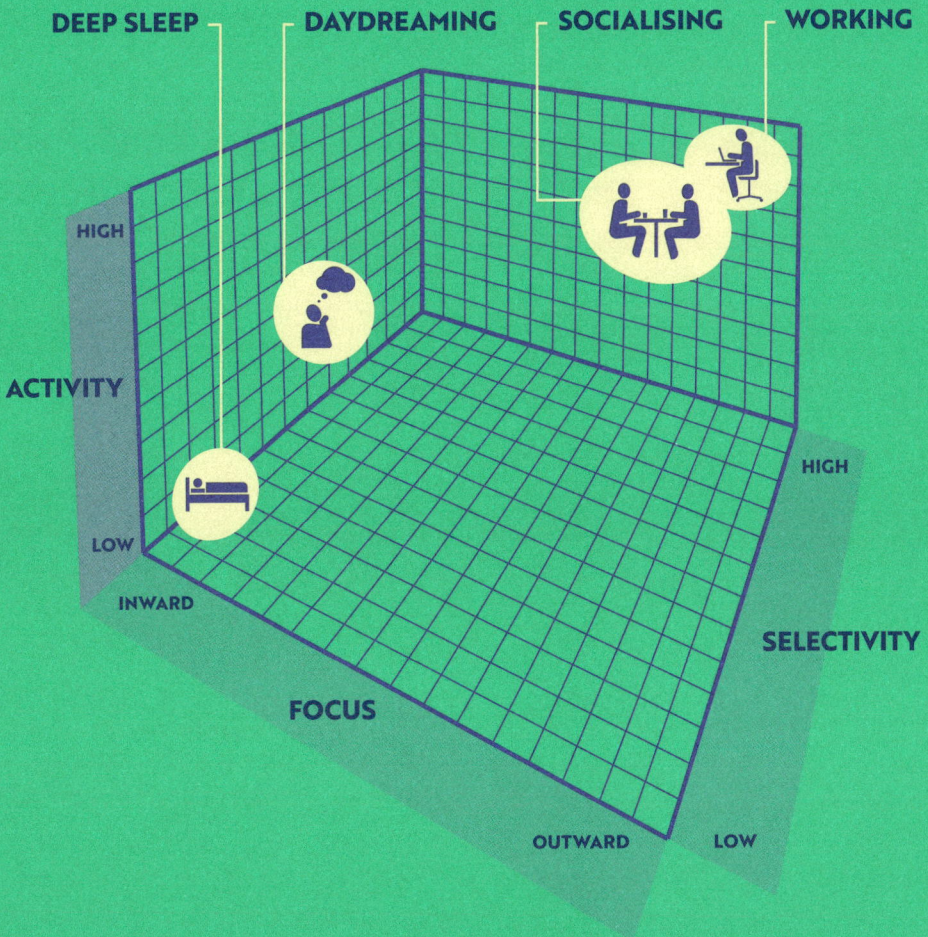

DEEP SLEEP   DAYDREAMING   SOCIALISING   WORKING

HIGH

ACTIVITY

LOW

INWARD

HIGH

SELECTIVITY

FOCUS

OUTWARD   LOW

Are you conscious of everything when awake?

⟶ **Consciousness can be likened to a spotlight that illuminates a certain area. It moves around and may be wide and blurry, or narrow and sharp. But it never takes in everything.**

⇗ Have you ever been baffled by one of those magic tricks in which objects disappear or change position under your nose? However closely you attend, the movement seems to happen magically. Ironically, it is 'close attention' that makes the trick possible. As the phrase suggests, it closes your attention to everything except what you think is important. The trickster guides you to look at all sorts of palaver while the really important thing – switching or moving the object – happens elsewhere. Magicians call it misdirection; psychologists call it inattention blindness.

The ancient art of misdirection wasn't given much scientific attention until the first years of this century, when studies showed that if people are told to attend closely to something, most will fail to notice other things, even when those things are in front of their eyes. Red circles on a computer screen go unseen when a person is told to attend to black squares; players stop slapping one another and shake hands, unnoticed by people told to concentrate on something a few feet away.

The same happens with sound: if a person is told to concentrate on one of two simultaneous conversations, all manner of scandalous tittle-tattle may pass them by. You can't always decide what to attend to, however. If, for example, your own name pops up in the conversation you are not concentrating on, your attention will switch to that conversation instantly. Although you were not conscious of the words until that moment, your brain will have been registering it unconsciously (as it registers many things around us) and your name grabbed the attention and yanked the entire conversation into conscious awareness.

'Change blindness' refers to a similar phenomenon: If you are looking at a moving scene, you may think you are conscious of everything in it. But if your view is blocked for a moment, huge changes in the scene may go unnoticed, demonstrating how little we really attend to.

CHANGE BLINDNESS EXPERIMENT

One change blindness study, by Harvard researchers, invited students to sign up for an experiment. They were told to get an application form from a man standing behind a counter. When they asked for the form, the man said he would fetch one, then went into an adjacent room, closing the door. Seconds later, a different person, wearing a different-coloured shirt, emerged with the form. Seventy-five per cent of the students failed to notice it was a different person.

Does your brain stop working when you sleep?

→ Your brain might slow down as you sleep, but thankfully the parts that maintain vital functions keep ticking over. Throw in some illusions, memory encoding, dreams and a spot of housekeeping, and it's clear there's no rest for this organ.

Sleep is divided into four stages: N1, 2 and 3 and REM (rapid eye movement, or dream sleep). Each one is characterised by different types of brainwaves. Brainwaves show how fast neurons are firing. They are measured using EEG, which typically displays them as horizontal waves or spikes, hence the name. We usually cycle through all four sleep stages every 90 minutes or so.

As you fall asleep your brainwaves slow down. Some parts slow before others, and those that remain firing fast may throw up brief illusions known as hypnagogic phenomena. Sometimes these replay something you did earlier. For example, Tetris – the computer game in which shapes float down the screen to be stacked neatly – figures so frequently in players' hypnagogic reports that the name of the game is now used in scientific literature to identify a certain type of illusion. Hypnagogic illusions usually fade as you enter N1, the lightest stage of sleep, marked by a mixture of alpha waves (eight to twelve per second). Most of us pass through this stage in five minutes or so.

In the next stage, N2, most cortical brain activity is slowed to between five and eight oscillations per second (theta waves), but occasionally there are bursts of faster, localised activity called sleep spindles. These help to encode memories in the cortex, where they may last for ever. The deepest stage of sleep, N3, consists of delta brainwaves that may be as slow as two per second. Although very little is going on consciously, you may still experience strong emotions. Night terrors, for example, occur in this state. Vivid dreams happen in REM sleep, when the brain becomes very active indeed.

When we emerge from a sleep cycle we may experience hypnopompic hallucinations. These are similar to hypnagogic phenomena but tend to be more coherent, as the brain is speeding up rather than slowing down.

METABOLITE CLEARANCE

Important body repair and maintenance takes place while we sleep, and one of the key housekeeping jobs is cleaning out the brain. Active brain cells produce metabolites – waste products caused by chemical reactions that happen as they fire.

These build up in the brain, contributing to the debris associated with certain types of dementia. When the brain slows down, in deep sleep, this detritus is washed out by cerebrospinal fluid. This may be the main reason that we need sleep.

Are dreams conscious?

→ Yes. When you're dreaming, you are as conscious as when you are awake. People may think they don't dream because they don't remember it, but that doesn't mean it doesn't happen.

Many people claim that they rarely or never dream. They are wrong. Practically everyone dreams several times every night, and the action in dreams is often dramatic, emotional and full of sensation. Dream experience is mirrored by the eruption of high activation in the brain, comparable to what would occur if the person was having the experience while awake.

Oddly, though, dreams are not encoded in memory as firmly as waking experience. The hippocampus, the brain area responsible for encoding and retrieving memories, is very active in dream sleep, but it seems to be engaged in retrieving recollections (dreams often weave in disjointed episodes from recent waking life) rather than encoding the current experience. It may also be helping memory consolidation, the shifting of recent waking experience from short-term storage in the hippocampus to permanent encoding in the cortex.

The brain does a fantastic job of creating a sensationally realistic dreamscape, but the events that take place in it are often bizarre and even impossible. People morph from one to another, items appear from nowhere and the narrative is often absurd. Yet in most dreams we accept it all unquestioningly, because brain areas that would normally alert us to oddities are disengaged during dream sleep. They include frontal lobe areas which in waking life keep us aware of who and where we are, and continuously (though unconsciously) pose the question: Is this really happening?

Occasionally these frontal areas wake up during a dream, while the hallucinatory dreamscape continues, along with the neurochemical block on motor instructions that keeps you paralysed during sleep and therefore unable to act out your dreams. The effect is to nudge you into a 'lucid' dream in which you can continue the dream as though in virtual reality and even control it.

SLEEP PARALYSIS

Lucid dreaming can be ecstatic, but if you don't know what's happening it can be terrifying. During sleep, signals from the motor area of the brain can be blocked, so you can't move. In this state it is common to feel a weight on top of you, as though someone is sitting on you. This is thought to be the origin of the incubus folklore – a demon that seeks to have sexual intercourse with sleeping women – and its female equivalent, the succubus.

What can create an altered state of consciousness?

⟶ 'Altered state' usually refers to some kind of ecstatic feeling outside the normal range of experience. There are many ways to get there, but safe and legal routes are notoriously hard work.

Any state in which the brain is functioning in an abnormal way is, technically, altered. Many illnesses bring changes in mind-state, as do injury, shock and, sometimes, extraordinary events. The feeling may be pleasant or the opposite.

Some altered states are ecstatic. They include, to various degrees, deep love, loss of ego or self, timelessness and joy. Such states are associated with stimulation of various neurotransmitter receptors, including those for glutamate (the brain's major excitatory chemical), dopamine (pleasurable anticipation) and serotonin (satisfaction). Occasionally people become ecstatic spontaneously or in response to a trigger such as music, sex or beauty. Most people, however, have to go to more unusual lengths.

The brain needs considerable encouragement to generate ecstasy, because making you feel good is not its purpose. It maintains our sense of space and time and self as structuring the world this way allows us to operate in it efficiently. It creates desire, dissatisfaction, competitiveness and longing – 'suffering' in Buddhist philosophy – to motivate us to find species-preserving needs such as food and sex.

For those who want to take time out from brute survival and bask in oceanic joy, there are ways to get there. The quickest – and riskiest – way is to take a drug such as an opioid (heroin or morphine), MDMA (ecstasy) or a hallucinogenic such as LSD or ketamine. The safe but difficult way is to take up and stick to an esoteric practice, such as meditation.

There are many schools of meditation – zen, transcendental, loving-kindness, mindfulness – and brain studies have shown different functional changes in each of them. A common factor between them is that they reduce the amount of time spent in what is known as 'default mode network' – a circuit that is activated in the brain when it is ruminating aimlessly, with one's self as the focus.

MEDITATION

Meditation usually needs to be practised rigorously to deliver the full benefit of altered states of consciousness. One 2015 joint study by researchers at the University of California and the Australian National University suggested it might slow aging by preventing cerebral atrophy – a loss of neurons and the connections between them – in many areas of the brain. Atrophy can occur in many brain disorders, including Alzheimer's.

Can AI replicate the human brain?

→ 'While AI has made significant advances in various domains, it is still far from replicating the full functionality of the human brain.' So says Chat GPT-3, which should know.

Chat GPT-3 went on to say that AI is 'based on neural networks that are inspired by the structure of the brain but are simplified abstractions'. Then it added that more sophisticated models are on the way. Would such models result in AI replicating the human brain?

AI is already close to being able to do all the intellectual tasks we can do, mostly better. We have devices in our pockets that can 'talk' in dozens of languages, do lightning-fast calculations and beat us at chess. Brain studies show that practically everything we do can be described in terms of information flow, so there seems good reason to suppose that our devices will soon be able to display emotion, predict and react to others' behaviour, come up with new ideas and create novel products. Our stored knowledge is dwarfed furthermore by AI's access to online data – Chat-GPT has 70 billion data points to draw on. But what about life experience and the thing that makes it all meaningful, consciousness?

In 1950 British mathematician Alan Turing came up with the Turing test. A human assessor interacts with other people and an AI system, unaware of which they are engaging with. A panel of judges assesses the conversation and if they cannot tell which one is with the artificially intelligent system, it passes the test. As yet no system has passed the test under the formal conditions stipulated by Turing.

Nevertheless, most researchers in the field believe that artificial consciousness will be with us soon, and some believe it is here already. Some think the plug should be pulled on further research, lest conscious AI usurp humans. Many people, including some AI researchers, even fear that digital intelligence could turn on its makers and kill us. Chat-GPT is very clear that it poses no such threat:

'I don't have a personal purpose or a life of my own ... ultimately the purpose I serve is defined by the goals and intentions of those who utilise my capabilities.' But, then, it would say that, wouldn't it?

THE CHINESE ROOM

Imagine yourself in a room in China. Locals slip written questions in their native language under the door. You don't speak Chinese but you have a database containing all the symbols, plus detailed instructions for manipulating them to arrive at appropriate answers. You slip these back to the locals. To them it seems that you, or maybe the room itself, is conscious. John Searle, the US philosopher who invented the Chinese room, thinks it is not – consciousness requires something more. Do you agree?

FURTHER EXPLORATION

BOOKS

Carter, Rita. *The Brain Book: An Illustrated Guide to its Structure, Functions, and Disorders.* Dorling Kindersley, 2019

Chalmers, David J. *The Conscious Mind: In Search of a Fundamental Theory.* Oxford University Press, 1997

Eagleman, David. *Livewired: The Inside Story of the Ever-Changing Brain.* Canongate, 2021

Gregory, Richard L. *Eye and Brain: The Psychology of Seeing,* Princeton University Press, 2015

Seth, Anil. *Being You: A New Science of Consciousness.* Faber & Faber, 2022

Taylor, Robert L. *The Deceptive Brain: Blame, Punishment and the Illusion of Choice.* Iff Books, 2021

ONLINE RESOURCES

BioPsychology NewsLink
biopsychology.com/news

Mind Hacks
mindhacks.com

Science Direct Brain Research
sciencedirect.com/journal/brain-research

The Brain from Top to Bottom
thebrain.mcgill.ca/flash/pop/pop_plan/
plan_a.html

BLOGS

Brainblog
neuropsychological.blogspot.com

NeuroBollocks!
neurobollocks.wordpress.com

Psyblog
www.spring.org.uk

Sharp Brains
sharpbrains.com

PODCASTS

Brain Stories
ucl.ac.uk/research/domains/neuroscience/
brain-stories-podcast

Hidden Brain
hiddenbrain.org

University of Oxford: Brain
podcasts.ox.ac.uk/keywords/brain

NOTES ON CONTRIBUTORS

AUTHOR

Rita Carter

A science and medical writer, lecturer and broadcaster, Rita Carter specialises in the human brain: what it does, how it does it and why. She has twice been awarded the Medical Journalists' Association prize for outstanding contribution to medical journalism and has been shortlisted for the Royal Society Prize for Science Books. Rita studied psychology at Oxford University and holds an honorary doctorate from Leuven University, Belgium, for services to the public understanding of the brain. theritacarter@gmail.com

ILLUSTRATOR

Robert Brandt

Based in the UK, for over 20 years Robert Brandt has been a visual communicator with a focus on illustrating technical and scientific subjects ranging from astrophysics to biochemistry. He works with experts to make complex topics accessible to a wide audience in publishing, industry and education.

AKNOWLEGEMENTS

UniPress Books would like to thank Robert Brandt for his illuminating illustrations and Luke Herriott for his elegant design work.

INDEX